CFZ
Yearbook
2022-3

Typeset by Jonathan Downes,
Proofed by Guinevere Palmer
Cover and Layout by SPiderKaT for CFZ Communications
Using Microsoft Word 2000, Microsoft Publisher 2000, Adobe Photoshop CS.

First published in Great Britain by CFZ Press

CFZ Press
Myrtle Cottage
Woolsery
Bideford
North Devon
EX39 5QR

ISBN: 978-1-909488-65-6

For Corinna

Introduction

Dear friends,

Welcome to the latest CFZ Yearbook. It has been a couple of years since the last one, but life, as John Lennon once said, is what happens to you when you are making other plans. For about 15 years I co-edited these volumes with my wife Corinna, but in July 2018 she was diagnosed with cancer, and despite a valiant battle over the next two years, she died in August 2020. And I'm sure you will understand that during her final illness all the other things had to take second place to my position as husband, and – bizarrely – as I had been kicked out of the national health service in 1990, a strange resurrection of my position within the nursing profession.

Her death was like a kick in the stomach to me and to everyone else who knew and loved her, and I spent a year or so trying to recover, and now I can fully take up the reins of leadership of the Centre for Fortean Zoology once again. The first thing that I had to do was to appoint another co-editor, and there was really only one candidate. As many of you know, I was a child in Hong Kong, which was then under British rule, and if I may quote from the volume of autobiography that I wrote covering those years:

"I spent the latter part of August and early September 1970 in hospital undergoing a barbaric operation on my knees. I went in on the day after my eleventh birthday, and stayed in for the next four weeks.

However, I was allowed to come home for weekends, and every Friday afternoon an ambulance would drive from Queen Mary Hospital down in Pokfulam, up the Peak, until I was carried into our ground floor flat at Peak Mansions by two burly young ambulance men. Somehow, I can surmise with the benefit of hindsight, that my father had pulled rank, because I had even been provided with a hospital bed on wheels, (something which I doubt was a service available to all and sundry) and on very sunny days Ah Tim and Ah Tam would wheel me out into the conservatory, and open the french windows so I could haul myself up in bed using the support bars, and look out onto the world outside.

Peak Mansions was (I use the past tense because it was apparently demolished in 1989) a six story squat building with an impressively mock Georgian façade, and two seemingly pointless green domes on the roof. It had originally been built, either late in the 19th or early in the 20th Century as accommodation for expat Civil Servants, but during WW2 it was the home of the Hong Kong Volunteer Force and was badly damaged by shelling. However it was seriously rebuilt after the war, and carried on with its original purpose. Running along the front of the building was Peak Road, and on the other side of the road was a heavily forested hillside which tumbled down for miles to the town (now city) of Pokfulam. This forest was the haunt of leopard cats, pangolins, civets, and porcupines, and within living history had been home to tigers and possibly even leopards. It was a magnificent place, and I spent much of my childhood exploring it, and much of my adulthood dreaming about it, and after a week of surgery (they botched the operation the first time and had to do it again) and physiotherapy, just to lie in my bed looking through the open french windows at the jungle below was bliss.

Outside the windows, a shallow sloping stretch of lawn led down to the road, and my mother was known to recline on a sun lounger there and sunbathe. Occasionally she would be joined by her friends, and on this particular occasion a lady called Sheila Muirhead, with an irritating young son aged four had come to visit. I was annoyed, as now my mother would not be willing to tell me stories, or make too much of a fuss of me, and what was worse, her son was too young for me to be able to talk to on any meaningful level, and as both my legs were in plaster, and I was wracked with agony every time I moved, I couldn't do anything more boisterous in terms of play.

Then I had an idea.

For my birthday, the day before I went into hospital I had received a copy of Hong Kong Butterflies by Major J.C.S Marsh, and I was desperate to put my newfound book to use. I was at the age when I had just begun to realize that some creatures were more closely related than others, and I wanted to identify the myriad animals that surrounded me. Fluttering along a few inches above the closely mown grass were dozens of small, blue butterflies. Major Marsh listed several dozen members of this family, quite a few of which looked very similar.

So I called to the toddler who was earnestly chuffing up and down the sloping lawn pretending to be a goods train.

"Hello" I said. "I'm Jonathan. What's your name".

"Richard". He said. "What are you doing in bed?"

So I told him, and despite the seven year gap in our ages, he not only seemed to sympathize with me, but – after I explained my predicament re. Major Marsh and the blue butterflies - he expressed - as well as a four-year-old can express anything – a willingness to help me in my investigations. So I told him where my bedroom was, and where I kept my butterfly net, and where my precious copy of Hong Kong Butterflies was, and he trotted off inside. About ten minutes later, after a few false starts, I was sitting up in bed with Major Marsh's magnum opus on my knee, and a Robinson's marmalade jar in one hand, as my young assistant – still making enthusiastic train noises – rushed up and down in search of butterflies.

When, eventually, he captured one, my quest was at an end. It was a fine male specimen of Lampides boeticus otherwise known as the long tailed blue. Having identified the poor little thing, Richard and I liberated it, and I thought nothing much more about the matter for about twenty-four years.

Nearly a quarter of a century later, whilst the nascent Centre for Fortean Zoology was in its infancy, I received a letter from a young man called Richard Muirhead, who was very interested in the stranger aspects of the natural world. It took months for me to make the connection, but eventually the penny dropped.

This time around, the seven year age gap was of no importance whatsoever, and we soon became firm friends. Thirty years later we still are."

Richard has what Charles Fort would have described as a `Wild Talent`. Like another old friend and quondam collaborator Nigel Wright, with whom I wrote a book called The Rising of the Moon ten years ago, he has a singular skill that I have always likened to that of a truffle hound; one of those elderly dogs that live in rural France that have the very useful talent of being able to sniff out the ridiculously expensive edible fruiting body of a group of subterranean ascomycete fungi of the genus Tuber.

Nigel and Richard have similar talents. Except it is not edible fungi for which they hunt. Both of them seem to have an almost supernatural ability to sniff out arcane information. And both of them are invaluable research tools. However, if you decide to utilise their peculiar abilities, one has to beware, because they will always end up nudging one's research into hitherto unsuspected areas.

Richard, has been a very valued colleague ever since the CFZ became anything apart from a figment of my imagination and my admiration of his academic skills, especially those involved with libraries and data retrieval knows no ends. His aforementioned 'wild talent' manifests itself in a myriad of peculiar ways. I will always remember how he told me in a jolly manner one day that he was going to look through the archives of the Hong Kong yacht club, I thought this was a stupid idea and a waste of time, but I should have know better… because, the next day he telephoned me to tell me of some seaserpent accounts that he had found within their archive. It had never occurred to me that he would be able to find anything of such importance in such a place. So, ever since, I have trusted his instincts implicitly and feel mildly guilty that there was a time that I didn't do so.

Therefore, when I found myself in the sad position of having to replace my late wife as collaborator in this volume of the yearbook, there really was only a short list of one name.

It seems that my faith in him has been very much justified, because right from the start he has thrown himself into working on this particular project with a gusto that one seldom sees in these decadent days, and I am very proud of everything he has achieved. Thank you very much, old friend!

I first came up of the idea of the yearbook back in the autumn of 1995, when - as he has just pointed out to me - my amanuensis Louis who is also the CFZ tech support and the person responsible for our magnificent new website, amongst other things, oh and producer of On The Track, was only three months old. I had always intended the yearbook to come out every year, but as I have already quoted John Lennon's famous line from 'Watching the Wheels' once in this editorial, I shall not repeat myself, but there have only been sixteen or seventeen volumes in the intervening years. This is largely because when one is dealing with such zoological esoterica as is the CFZ, the amount of subject matter for potential inclusion into a volume like this one is, perforce, limited. I am very proud of the quality of the entries in all our yearbooks and would rather keep this quality at the highest level and not sacrifice it in favor of having to produce a yearly publication. And there is, of course, John Lennon's aphorism about life to consider.

And then, on top of all this I was in my mid thirties when we started the CFZ yearbooks and I am now in my mid sixties and I am not in particularly good health. So, as a result of all that, this yearbook is a biannual publication (this does not mean anything sordid as implied by my increasingly ridiculous amanuensis, of whom I am actually very fond, despite the fact that we argue continually like female pop stars trying to get into the news pages). Note that Louis is someone I expect you will be hearing more from soon, as I prepare to hand the CFZ over to the next generation.

I hope you enjoy this issue and that it may propel you into getting more involved with the Centre for Fortean Zoology than you are already.

Yours, as always

Jon Downes
Woolsery,
February 2022

Contents

KIWI.

On the Track of Smaller Hidden Animals
The Great Days of Microcryptozoology Have Just Begun

Carl P Marshall

Many cryptozoologists will outright reject reports concerning small unknown animals, considering them unworthy of attention. Dr Bernard Heuvelmans, the "Father of Cryptozoology", insisted that in cryptozoology a minimum size is essential, but unfortunately, never committed himself further. I suspect his reluctance was simple. Other than professionals, far fewer observers would bother reporting unusual butterflies, lizards, or dull looking fish the size of tetras.

Smaller unknown animals typically have fewer eyewitness reports betraying their biological identities, thus requiring greater consideration by investigators. This lack of broad recognition along with a few other subtle distinctions, demands a revised and amended approach when dealing with data respecting diminutive unknown animal forms. Unlike traditional cryptozoology, where it's very much a case of the more reports the better, in microcryptozoological studies, qualitative data typically has a more productive role than a simple numbers perspective.

This rejection of smaller, apparently less sensational unknown animals has not strengthened cryptozoology's already rocky relationship with conventional science. Despite its recent surge in popularity, made possible in part by social networking websites like Facebook, Twitter, WhatsApp, and Instagram, cryptozoology has deteriorated from a nascent zoological discipline into something more akin with paranormal research and mysticism. Therefore, by embracing the concept that small unknown animals are not just relevant but are in fact an integral and highly viable part of practicable cryptozoology, cryptozoologists will be providing potential insights for many more future zoological discoveries. By utilising but adapting Heuvelmans' methodology to target the potential plethora of microcryptozoological animals, we can dramatically improve cryptozoology's overall rate of success and subsequent acceptance as a legitimate field of study, if not actually a branch of zoology, by the conventional scientific community. Microcryptozoology, like studious cryptozoology, is a legitimate field of study and capable of providing regular results.

To my knowledge, the term "micro-crypto-zoology" (directly explained as unknown or hidden animals of small size) was first suggested by biologists Aaron M. Bauer and Anthony P. Russell in 1988, and later independently by Danish zoologist Lars Thomas, and at present *loosely* describes the evaluation of evidence for unknown animals of small sizes.

Cryptozoology, if done properly, is the unbiased evaluation of evidence for unknown animal forms. Evidence is either anecdotal, testimonial, circumstantial, and very occasionally biological in the form of samples of unidentified zoological material. Cryptozoology is not a science, not in the way that entomology is a branch of zoology primarily concerned with insects, or paleontology is the study of ancient life through fossils. Cryptozoology is, in my opinion, best described as a scientifically driven discipline dedicated to the study of mystery animals and animal mysteries. Foremost it's a targeting methodology.

Cryptozoology enthusiasts regularly assert that "cryptozoology becomes zoology" following formal zoological discoveries. I propose that cryptozoologists working in the field, whether misguided of not, are in fact prospective zoologists as soon as field research for unknown animals begins. The achievability of our fieldwork reflects either the presumed rarity and/or unexpected nature of such secretive animals, the limited time, and resources available for field expeditions, difficult and often inaccessible, unpredictable landscapes, or, whether or not we care to admit it, our erroneous interpretations of the available data. The only aspect of fieldwork that can perhaps be considered 'cryptozoological' is the acquisition and assessment of fresh testimonial and anecdotal evidence carried out in the natural environment.

Bernard Heuvelmans essentially presented a methodology that can be used to target and subsequently identify new species, however, it's up to us as cryptozoologists how we interpret and utilise data. Cryptozoology is the process; the outcome is either zoological or a comprehensive revaluation of the evidence is necessary. Cryptozoologists should stop regurgitating and repeating defunct data. Simply put, if the investigation is properly executed, meaning if an investigator is a competent zoologist or naturalist, wildlife tracker or field naturalist, following strict scientific principles, cryptozoology bridges uncomplicatedly into zoology and successes are achieved.

Heuvelmans' ultimate classification of cryptozoology provides the fundamental framework employed and described in this current work, which can be thought of simply as a necessary and long overdue supplement to the cryptozoology methodology.

One element of cryptozoology, and therefore also of microcryptozoology, is that the animals studied should be *"ethnoknown"*, meaning if discovered and validated, a true cryptid should be previously known to the local people who share its habitat and to some extent have interactions with it. These witnesses then typically go on to describe the unknown animal to visiting intermediaries who will then hopefully make the data available in the relevant publications.

For a newly described species to be formally acknowledged as a cryptozoological success there must be:

(A) **Some prior knowledge** of the animal's existence before its formal discovery and taxonomy (i.e., ethnoknowledge or perhaps known by unclassified biological materials etc.), and

(B) **It must be, above all else** characterised by some unexpected trait[s], either in terms of its morphology, its physiological functions, behavioural traits, or perhaps simply its presence at a particular time or place.

The progenitor of the cryptozoological process is ambiguity! It is the ambiguous nature of cryptids that sets them apart, making them notable. However, ambiguity is a characteristic that cannot be precisely measured. An aspect perceived to be out of the ordinary to one observer might be quite normal to another. Ambiguity is therefore open to interpretation. It should be standardised however, that in microcryptozoology, there *must* be some unusual characteristic[s] that noticeably sets an unknown animal apart from catalogued species known to inhabit the same location.
Does this mean then, that if a trained observer such as an entomologist happens upon an unfamiliar insect in the field and due to vexatious

circumstances is unable to provide a definitive type specimen and convince their peers of its existence, that this unusual insect is not of cryptozoological importance? Of course not! This is an incredulous and counterproductive proposal, especially within micro-cryptozoology, where it is, in fact, a reasonably commonplace occurrence.

There are literally millions of species, mainly invertebrates, that have not yet been scientifically described and most of them will be discovered and validated without any prior knowledge of their existence, and are therefore, by definition, not true microcryptids. Over 1.5 million living animal species have been described, of which approximately one million are insects, but it has been estimated there are over seven million species in total (some sources even suggest as high as thirty million). Smaller organisms, in terms of probabilities alone, have far greater chances of discovery than larger organisms in any environment. We often hear critics of cryptozoology inquiring why haven't any self-acknowledged cryptozoologists ever discovered the mystery animals they seek? This is a valid point which deserves to be addressed. Cryptozoologists have admittedly had very little, if any success in the field. This unfortunate fact, is, I believe, mainly because cryptozoologists habitually focus their efforts only on the more fantastical aspects of cryptozoology.

Unfortunately for analytical cryptozoology and for science, the legitimate search for unknown animals has become entwined in a mix of the mystical and pseudoscientific, further blurring the boundary between myth and reality.

In truth, serious cryptozoologists may as well be searching for moving needles, concealed in a proverbial field of haystacks, and as we cannot know precisely which "needles" (targets) are genuine and which are simply the "hay" (hearsay), we must objectively, not subjectively, investigate *all* the available evidence.

So why then are cryptozoologists largely ignoring these smaller unknown animal forms when their greater potential for discovery can only strengthen the case for the few remaining large ones? And not to mention the scientific vindication of the cryptozoological methodology in general. I also expect this will help in highlighting the struggles discovering the more biologically feasible traditional cryptids, such as the purported ape-like Orang Pendek and the supposedly extinct thylacine *(Thylacinus cynocephalus)* also commonly known as the Tasmanian tiger or wolf.

This species was once native to the Australian mainland (until about 2000 years ago) and the islands of Tasmania (into the 1930s [*]) and New Guinea (end of the Pleistocene), the thylacine is now, I believe, most likely to be found alive (if they have survived anywhere) in Indonesian New Guinea based firmly on ecological data rather than the more traditional cryptozoological lines of enquiry. My number one rule as a cryptozoologist is *ecological forecasting first, cryptozoology second!*

Large aquatic cryptids aside, I expect there are still numerous decent sized terrestrial animals left to discover. Maybe even a few 'true giants' hidden away in the last truly isolated regions of our planet, but by far, the largest cryptozoological haul is potentially that of the microcryptids, which have the very real possibility of being discovered the world over and may even haunt one's own back yard!

Officially cryptozoology is not universally accepted as a valid scientific field of study and is unfortunately all too often considered to be simply crackpot pseudoscience, or worse still "monster hunting". As a working cryptozoologist with a background in entomology, I consider this negative and unfortunately popular view of pseudoscience as unjustified. It couldn't, or at least it *shouldn't* be further from the truth. The analytical cryptozoologist isn't thinking about monsters, they are attempting to discover and validate the unknown animal behind the myth, behind the monster. If indeed there is one.

'Monster' is simply a term we use for unsubstantiated creatures or organisms we inherently fear. The yeti is only a monster while it remains a mystery. If one is discovered tomorrow as a flesh and blood biological species, it would simply become a new mammal (albeit a very exciting one) in the journals of zoology, and at least in theory, a totally new higher primate species.

Microcryptozoology cannot be considered monster hunting, although some readers might consider the potential for glowing spiders or bizarre new snakes, rats, and bats, to be monstrous enough. Monster comes from the Latin *monstrum*, itself derived from the verb *moneo* ("to warn, or foretell"), and denotes anything "strange or singular, contrary to the typical

---

[*] A 2021 paper *Extinction of the Thylacine* by Barry W. Brook1,2*, Stephen R. Sleightholme3 , Cameron R. Campbell4 , Ivan Jarić5,6 3 and Jessie C. Buettel1,2 4 . Is a mainstream paper proposing that the species survived at least into the 21st Century.
https://www.biorxiv.org/content/10.1101/2021.01.18.427214v1.full.pdf

course of nature", "a strange, unnatural, hideous person, animal, or thing," or any "monstrous or unusual thing, circumstance, or adventure."

The definition in the *Oxford English Dictionary* denotes the term "monster" as meaning:

1)   A large, ugly, and frightening imaginary creature.
       1.1) an inhumanly cruel or wicked person.
       1.2) a rude or badly-behaved person, typically a child.
2) a thing of extraordinary or daunting size. [or]
3) a congenitally malformed or mutant animal or plant.

Of these definitions, only the original Latin does any justice to the intellectual scope of cryptozoology, and those of us who feel misrepresented by the state of contemporary cryptozoology and who consider a cryptid's dimensions *not* to be the deciding factor of its being, should unite to address this major issue facing cryptozoology by adopting and embracing and a viable microcryptozoological methodology.

Monsters are typically hideous, and they are typically large. We are told that large size is strength and size is power. Large equals fitness - the biggest survive, and that size has influence and is value for money. The Western world is urged to respect size and ignore quality. But for this, our attitude towards size is equivocal; we both fear and admire it. On the one hand we are delighted when large size is overthrown, be it a man putting a monster in its place such as St George and the Dragon, the small female overcoming the big male as in Sampson and Delilah, or simply the small and good overcoming the big and bad as told in the biblical story of David and Goliath. It gives us great pleasure to see financial tycoons topple, bosses dismissed, commercial empires dissolve, World Cup winners beaten, and boxing champions KO'd. On the other hand, we seem to impulsively admire anything big: the tallest buildings and the great ocean liners attract our admiration, and, hopefully, the large animals such as the elephants, the giraffe, and whales – our compassion.

Finally, and fundamentally, monsters must be undiscovered. One might say discovery utterly discredits their elusive reputations! The case of the largest known primate in the world, *Gorilla gorilla*, is a wonderful and noteworthy example of the potential fruits of practicable cryptozoology. For over two hundred years until its discovery in 1847, these mighty primates were considered nothing but made-up monsters - figments of the uneducated native people's overactive imaginations. Not real blood and bone animals.

(Note the study of unknown animals was then usually known as 'romantic zoology', or 'exotic zoology', and that the terms 'cryptozoology' and 'cryptid' were yet to be coined).

In recent years, Western investigators have come to realise the error of their previous thinking about native peoples. Studies have shown many apparently outlandish-sounding claims by native observers to be true, or at least to have a basis in fact. A noteworthy example is the assertions by New Guinean tribesmen that certain local birds are poisonous.

The gorilla is the perfect prototype for all problematic cryptozoological 'man-beasts', as, in many respects, this poor, largely inoffensive primate was once maligned with many of the same negative folkloric behavioural traits, or at least ones very similar, which today are still attributed to the so -called Himalayan Snowman - the yeti. We now know that if unprovoked, the gorilla is all but harmless to humans, a gentle giant (in stark contrast with the malicious nature chimpanzees, which are incidentally, and somewhat predictably, our closest living relatives).

The mighty gorilla isn't a monster and—of course—it never was.

My intention in writing this somewhat challenging paper is to aid in dismissing the prevailing views and to hopefully steer cryptozoology ever closer toward its zoological matriarch, and perhaps optimistically, helping it achieve the universal academic acceptance it long deserves.

Cryptozoology is first and foremost, the study of unconfirmed animal forms. The existence of which is based on

(1) testimonial evidence or
(2) circumstantial evidence, or
(3) material evidence typically judged insufficient.

Incidentally, zoologists and wildlife biologists often use aspects of cryptozoology when attempting to locate new species for taxonomic classification, yet they rarely, if ever, recognise it as such. A current example would be popular field biologist Forest Galante, who has successfully adopted certain aspects of the cryptozoological methodology to much fame and fortune. One might say he's a cryptozoologist by any other name and getting rich doing so!

There is more than enough space within cryptozoology for the microcryptids (no pun intended).

The question that naturally arises now is how small should an unknown animal be in order to be considered microcryptozoological? 'Micro' essentially means *very small*, and often relates to organisms too small to be observed with the naked eye – i.e., microscopic. This definition is all but useless in cryptozoological investigations as truly microscopic organisms are unlikely to ever be ethno-known, as it takes reasonably advanced technology and knowledge to observe and identify such tiny creatures.

Micro-cryptozoological animals *must be clearly observable to the naked eye* without the assistance of any technological visual enhancement, other than perhaps spectacles or contact lenses.

Some cryptozoological investigators suggest that microcryptozoology is the study of unknown micro-organisms. Technically, this would be called *micro-crypto-biology* because most microscopic organisms do not belong to the taxonomic kingdom Animalia – the animals. Animals (also referred to as metazoa) are multicellular eukaryotic organisms. With few exceptions, all animals consume organic material, breath oxygen, can move, can reproduce sexually, and grow from a hollow sphere of cells called the blastula during embryonic development. If an unknown organism is not expected to belong to the kingdom Animalia, it isn't cryptozoology, simple!

Micro-animals which have been formally described but are now contested due to the loss of their type specimens are sometimes erroneously regarded as microcryptids and microcryptozoology. They are nothing of the sort.

Testimonial, circumstantial, and obscure material evidence have previously led to discoveries of unknown species, sometimes spectacularly so, clearly showing cryptozoology's potential as a scientifically driven zoological discipline.

Evidence can be indirect; obtained from archaeological relics, artwork, old manuscripts, or legends and oral traditions, or direct; in the form of disputable material evidence, or verbal or written accounts of observations of supposedly undescribed animals from people either local to an area, or from outsiders such as explorers or travellers.

While attempting to codify microcryptozoology, I promptly realised it would be practically impossible to provide accurate dimensions in terms of the expected volume of an alleged cryptid, i.e., the amount of space that a substance or object occupies. Since the testimonial and circumstantial evidence required for cryptozoological investigations do not typically allow for such precise measurements, an additional approach would be to provide a maximum weight that the relevant unknown animals *presumably* weigh less than.

After much deliberation (and without revealing too much before the official publication of my main work on the subject), I suggest that a *Presumed Maximum Weight* (PMW) for microcryptozoological animals must be no greater than 4.4 pounds (two kilograms - approximately the weight of a ferret).

We can never be certain as to the precise measurements of an unknown animal, much in the same way we can't be certain of one's biological identity without the sufficient biological material for a detailed study (much to the disappointment of many investigators who claim to be cryptozoologists and believe that even without sufficient evidence, they already know exactly what type of creature they are searching for). At the very best all we can do is objectively assess what information we do have and provide the most reasonable solution based on that. Essentially, a well-informed guess.

The fundamental fact that new discoveries of large animals are exceptionally rare, obviously means that many microcryptids are likely to be small terrestrial invertebrates such as insects and arachnids. These creatures without vertebrae are by far the most abundant lifeforms on earth. Diminutive vertebrates such as small mammals, like rodents, and appropriate sized fishes, amphibians, reptiles, and birds also make up the envisaged Lilliputian fauna of microcryptozoology. Microcryptozoology, as it is, utilises the precepts of zoology and evolutionary biology, as does practicable cryptozoology.

Microcryptozoology is part of the broader field of cryptozoology, and is, in my opinion, at least applicable to conventional zoology and should be appreciated as such and can be defined as the systematic study of unknown animal forms presumed to weigh less than 2 kg (4.4 pounds) based on the data derived from testimonial, circumstantial, and/or disputable biological materials.

These unknown animals must be clearly observable to the naked eye. If an alleged microcryptid is quadrupedal, it should be no higher than 25 cm (250 mm) to its haunches and be no more than 1 m in length (head to tail tip). This length may seem excessive for species estimated to weigh less than 2 kg, but this allows more room for error and the alleged cryptid is still noticeably small. If a microcryptid happens to be bipedal it should be no taller than 40 cm (400 mm) total hight. This small height limit is due to the necessary mass required to support a bipedal frame and the density that accompanies it, and the subsequent weight that obviously accompanies these stresses.

This admittedly only allows for the very smallest of the alleged crypto primates on record, with undescribed simians (monkeys and small apes) likely being the most appropriate forms. These characteristics are not set in stone however, as one can never be certain what might in fact still be out there awaiting discovery.

All microcryptids will presumably weigh less than 2 kg (4.4 pounds), be no more than one metre in length, and no taller than 40 cm. Admittedly, the systematics of this classification scheme rely to some extent on a specific Class by Class premise. Certain animals, particularly birds, often vary greatly between size and weight and therefore should be considered accordingly. The wingspans of birds, bats, insects, and potential flying reptiles are also crucial and should be considered accordingly.

A comprehensive microcryptozoological biography will be presented in my forthcoming work on the subject.

*****

Here, exclusively for *Animals & Men*, we will begin by discussing two unique spiders pertinent to the microcryptozoological methodology. We start with a newly described species - an ex-microcryptid if you will.

Some Singular Spiders
In the light of a crackling campfire in south-eastern Angola, entomologist John Midgley, examined a strange-looking tarantula he captured. Midgley knew he had found a once in a lifetime discovery. A large, flexible horn sat

squarely on the arachnids back. This bizarre horned arachnid is completely new to zoologists, but well-known to the region's local population, who call it *"chundachuly"*.

Midgley is not an arachnologist, so he messaged photographs of his find to his collaborator, Ian Engelbrecht at the University of Pretoria in South Africa.

"Ian accused me of photoshopping the pictures," joked Midgley, of the *KwaZulu-Natal Museum in South Africa.* He went out the following night and found several more tarantulas with the same fleshy horn to prove his find.

"I knew then we had discovered a new species [though not new to the local inhabitants]. *It's rare to know you have something special so early in the process,"* he said.

The team named the new tarantula *Ceratogyrus attonitifer*, from the Latin for "bearer of astonishment," and published their results in the scientific journal *African Invertebrates*. Following a 26-year civil war that ended in 2002, Angola's biodiversity largely remained a mystery and no one honestly knew how many species had survived. In 2015, the *National Geographic Society* and an international team of scientists launched the *Okavango Wilderness Project* to survey and protect this important and underappreciated region. The project invited several experts, including Midgley, to central and eastern Angola to discover what species lived there.

In November 2016, Midgley was traversing Angola looking for insects, as well as spiders, scanning the ground for signs of his multi-legged friends. In a grassy seasonal wetland surrounding a lake in Angola (Midgley did not disclose the exact location to prevent the theft of these new tarantulas for the lucrative pet trade), he identified a series of inch-wide holes going almost two feet straight into the ground. Checking if anything was inside, he inserted a blade of grass into the hole. Immediately something tugged on the end. He returned that night, and as soon as he felt a bite at the other end, he slowly pulled the tarantula from its burrow.

"It was a lot like fishing," said Midgley. *"If you don't hold on tight, they can pull the grass right out of your hand."*

The large horn on the spider's back immediately classified it as a member of the genus *Ceratogyrus*, but this one was very different from anything heretofore documented.

Many spiders in this group have similar protuberances, but they are much smaller and solid. The flap on the back of *C. attonitifer* is the same length as its abdomen and is fatty rather than solid. It almost looks like an extra leg protruding from the spider's back, or perhaps it resembles a spider with some weird, hairy, worm-like creature riding upon its back or attacking it! Its potential for folkloric embellishments should be obvious.

Scientists know very little about this new spider, including how it uses its strange horn. They do know that *C. attonitifer* is a nocturnal ambush predator, sleeping deep within its burrow during the day, while spending nights at the entrance just waiting to pounce on insects and other unexpecting prey.

Like all spiders, *C. attonitifer* uses venom to kill and dissolve its victims, digesting the nutritious insect "soup" after its venom has taken effect.

Although Midgley found ten burrows in the 300 square meters surrounding his campsite, a high density for a predator, he only found the spiders surrounding one Angolan lake. Baboon spiders are very particular about where they live.

"Baboon spider" is a generic term for a subfamily of tarantula endemic to Africa, known to arachnologists and entomologists as Harpactirinae. One species might make its burrow only in one type of soil, another might build next to a particular type of rock. If their habitat is disturbed, these spiders can't just relocate. This, combined with their long-life spans and low reproductive rate, makes this newly discovered species highly vulnerable.

As well as providing microcryptozoological intrigue, this kind of basic biodiversity research goes a long way to solving the mysteries of the Okavango.

Known as "chundachuly" in the Luchazi language, the species is reported to prey predominantly on insects. The venom is not considered to be particularly dangerous, though, like with many species bites may result in secondary infections which can be fatal due to poor medical access. The subfamily is after all well known for being aggressive. It is claimed by the locals that the females take over and enlarge existing burrows rather than digging their own, though this has yet to be confirmed as both behaviours are known in harpactirines.

The new species of *Ceratogyrus* described here is truly remarkable. No other spider in the world possesses a similar foveal protuberance and the function of this strange horn is currently uncertain.

This enigmatic tarantula, known previously to the local inhabitants as a distinct type due to its strikingly singular anatomy, is certainly very exciting and indeed thought provoking. The problem is that cryptozoologists, overall, do not specifically search for these kinds of animals. Therefore, discoveries like this are never associated with cryptozoology. Thanks to programs like *Discovering Bigfoot* and *Mountain Monsters*, cryptozoologists are these days typically considered "monster hunters" in the eyes of most scientists and laypersons alike. And quite understandably so, considering these television programs have basically all but replaced the literature and are now most people's only exposure to the topic.

Had the local inhabitants been consulted on the regions more unusual invertebrates prior to the expedition, a cryptozoological search for the chundachuly would have provided a fantastic and exciting microcryptozoological exercise.

Staying for now with our eight-legged friends, we will now consider an unconfirmed bio-luminous spider, a true microcryptid, reported only once in 1923 from Central Burma (now Myanmar) by famed American Palaeontologist, and discoverer of *T. rex*, Barnum Brown (b. Feb 12, 1873 – d. Feb 05, 1963).

Brown's report in full:

One day in Central Burma the trail in the jungle was exceptionally difficult. It was long past noon when I realized that the return journey would be equally long and tiring. Camp lay on the other side of a long range of hills and there was a shortcut from the main trail that would save several miles, but this trail was faint. I reached the supposed cut-off about dusk and followed it upward.

Darkness came on swiftly and my pony began to stumble. Somewhere we had missed the trail, for at intervals I could glimpse the crest of the hills and I knew my general direction. Fireflies sparkled here and there.

Presently a few feet away I saw a ball of light as large as one's thumb. It was stationary. Tying the horse, I approached it as carefully as possible, finding it surrounded by thorny bushes. It did not move, and I pressed the brush aside until I was directly over it and then struck a match. There in full view was a spider, his large oval abdomen greyish with darker markings. Still, he did not move, and as the match died out his abdomen again glowed to full power, a completely oval light, similar in quality to that of the fireflies. Remembering native tales of poisonous insects, I wrapped a handkerchief around one hand, parted the brush with the other, and when close enough made a quick grab.

Alas! The handkerchief caught on a stick before I could encircle him, and my treasure scurried away. I followed as quickly as possible, but the light soon disappeared under stones, brush, or in some burrow, for I never saw it again.

Many nights I searched in the jungle and questioned natives and white officers who had passed through the district, but apparently no one else had reported a luminous spider, nor can I find record of any known elsewhere. Burmese never leave their homes after dark on account of their fear of spirits, so it's not surprising that the natives had never seen one, but some other traveller may be so fortunate as to capture one of these spiders.

The place where I saw the specimen was between the villages of Kyawdaw and Thitkydiangi, Pakkoku District, about one hundred and twenty miles west of Mandalay, Burma, in April 1923.

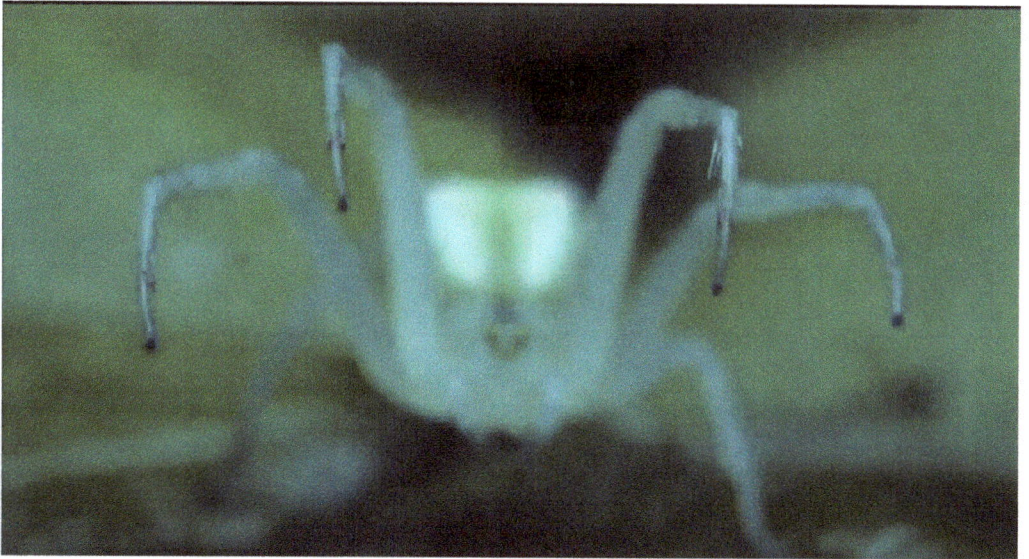

Barnum Brown, *American Museum of Natural History*.

If the ingestion of photo-luminescent (PL) material, such as the light-emitting organs found in fireflies was indeed the cause of the spider's glow, it is tempting to conclude that an ambush predator such as a spider, might benefit considerably by consuming bioluminescent insects. It makes perfect sense that some spiders might purposefully hunt fireflies or glow worms, if only for the simple reasons of visual stimulation and energy efficiency. These insects are easily observable, literally glowing in the dark to the typically nocturnal eight-legged hunters. If this behaviour proved beneficial by regularly attracting large nocturnal insects for predation, it could potentially become a frequent hunting technique under the right circumstances, in suitable ecosystems, especially for those species which do not produce webs.

Perhaps we have simply never noticed this cryptic behaviour before!

Given the solid oval light confined to the entire abdomen described by Barnum, direct bodily contact with bioluminescent fungi is an unlikely explanation in my opinion. It is far more probable that Barnum's glowing spider had previously ingested PL material.

The crab spiders first described in 1875 by Ludwig Carl Christian Koch, might provide a possible explanation. One species, the Australian *Tharrhalea* (formerly *Digea*), whose range extends from South-East Queensland to at least Sydney in New South Wales, was recently filmed by an amateur naturalist glowing at night.

I managed to locate the footage (see link below), which I promptly forwarded to colleagues from the *Stratford Upon Avon Butterfly Farm*, and to freelance arachnologist and CFZ Oxfordshire regional representative, Carl Portman, for their professional opinions. They each replied that the video looks genuine enough and agreed that it was probably caused by the ingestion of PL material, although they had never seen or heard of this behaviour in arachnids before. In the video, the markings observable on the dorsal surface of the spider's abdomen seem to correlate with the pattern and contrast of the bioluminosity. This might correspond with the sac-like diverticula of their digestive tract.

(https://www.youtube.com/watch?v=mGhW5usURNM)

Individuals of the closely related genus *Misumena* can change colour over a period of a few days to match the flower or leaves on which they are sitting, where they await possible prey. Some of them even sit out in the open. In my opinion, it is plausible that a few species might also utilise PL as a potential luring strategy.

Even though I would love to seriously suggest that Barnum Brown's glowing spider was nonother than a hitherto unknown, truly bioluminescent species endemic to Central Burma, this is probably less than likely. Any alleged species with no known contemporaries alive today or known from the fossil record must be treated with caution. It is unwise however in zoology to jump to premature conclusions, I therefore shall remain prudent even when contemplating this somewhat less than likely option. It should be noted however, that the presence of soft, highly specialised organic structures, such as those found in truly bio luminous species very rarely show up in fossils.

Bioluminescence occurs when an organism converts chemical energy into light energy. It can also occur because of bacterial or fungal activity, such as the glow from foxfire (aka fairy fire or chimpanzee fire). Bioluminescence serves many purposes, from communication to finding mates, scaring off potential predators to attracting prey. While many marine species use bioluminescence, very few terrestrial animals have evolved the ability. Besides fireflies and a few other insects, only one snail, a few earthworms and a handful of millipedes can produce light. There is now also additional evidence indicating that bio luminous organs evolved much more recently in terrestrial species compared with marine organisms. As evolution is a dynamic process, surely such an evolutionary adaptation must at least be possible.

Other than scorpions, which glow a bright cyan green under ultraviolet light, there are no truly bioluminescent arachnids known from anywhere in the world; therefore, it would be quite unexpected if Barnum Brown's glowing spider turned out to be a first.

Barnum Brown may not have been an entomologist, but he was certainly a trained and competent observer. Without doubt, what he briefly observed that night in Central Burma, was none other than a true spider emitting some sort of light. There is no reason to suspect otherwise.

As already mentioned, the snail *Quantula striata* (aka *Dyakia striata*) is completely unique among terrestrial gastropods in that it is bioluminescent. It's eggs glow in the dark and the juveniles, as well as most adults, give off flashes of green light. There are very few bioluminescent cockroaches, but they do exist. One such species is *Lucihormetica luckae*, they have spots on the carapace that glow when exposed to light (autofluorescence), perhaps to mimic the appearance of the toxic click beetle (*Pyrophorus*) that emits light at the same wavelength. This would then be an instance of *Batesian mimicry*. Autofluorescence is the natural emission of light by biological structures such as mitochondria and lysosomes when they absorb light and re-emit it upon excitation. Each spot is filled with bacteria that glows when exposed to fluorescent light. The effect is intensified by the fact the spots are covered with a reflective surface, making them act like a car's headlights. There is also anecdotal evidence suggesting that *L. luckae* is genuinely bioluminescent, though this is still inconclusive.

So why then not a spider?

Might Barnum Brown's spider have likewise been autofluorescent? This seems unlikely as the report clearly states that the light emitted grew dimmer when Brown struck a match. When the match eventually died, the light coming from the spider once again glowed to full power and not the other way around, as one might expect had this been caused by the re-emission of fluorescent light. It sounds more like either PL, or a hitherto unknown spider displaying a genuine bioluminescent organ.

Whatever the solution, Brown's report is undeniably credible and should be taken seriously. Might it be an accurate yet previously unrecorded observation of PL in arachnids? Unfortunately, we may never know for certain, since the location of Brown's brief observation has been hit hard in recent years by deforestation companies, meaning the species is now quite possibly extinct.

However, all might not be lost! Some miles to the west of Pakkoku District lies *Nat Ma Taung National Park* which covers a massive 713.06 km^2. The colossal Mount Victoria might also provide a suitable refuge for such an unusual and seemingly secretive spider.

Hopefully the *Glowing Spider of Burma*, or a similar, closely related species, might have survived there undetected to the present day.

Brown stated that the local people were unaware of this glowing spider, which he suggested might be because of their superstitious beliefs and the nocturnal activities of the spider. That said, just because Brown did not find anyone who had personally observed such a spider in the short time he was there, does not mean one has never been seen. Absence of evidence doesn't necessarily mean evidence of absence! Might a series of microcryptozoological field investigations into Central Myanmar reveal further data, perhaps even the mysterious glowing spider itself?

*****

Finally, we have two uniquely formed ethno-known mammals, one belonging to the family Soricidae - a shrew, and the other a rodent. Both might have remained undetected had it not been for the balanced use of ethno-known data.

Known to the people from Vangunu Island in the Solomon Islands as the vika, this large arboreal, hedgehog-sized rat is credited with the Herculean ability to crack open coconut shells with its teeth. Mammologist Tyrone Lavery first heard rumours of this incredible rodent on his first trip to the Solomon Islands in 2010, officially making its subsequent discovery a cryptozoological success story, even if it's not an obvious one.

"When I first met with the people from Vangunu Island in the Solomons, they told me about a rat native to the island that they called Vika, which lived in the trees", proclaimed Lavery. *"I was excited because I had just started my Ph.D., and I'd read lots of books about people who go on adventures and discover new species."*

However, after years of searching, and in a frustratingly familiar cryptozoological fashion, the investigation didn't turn up any of the coconut cracking rats.

"I started to question if it really was a separate species, or if people were just calling regular black rats 'vika'".

Part of what made the search for this creature so difficult was the rat's arboreal lifestyle.

"If you're looking for something that lives on the ground, you're only looking in two dimensions, left to right and forward and backward. If you're looking for something that can live in 30-foot-tall trees, then there's a whole new dimension that you need to search," explained Lavery.

Then finally, after seven arduous years of fruitless searching, one of the rats was finally discovered scurrying out of a felled tree.

"As soon as I examined the specimen, I knew it was something different... There are only eight known species of native rat from the Solomon Islands, and after looking at the features, I could rule out a bunch of species right away".

After carefully comparing the specimen with similar species in museum collections and checking the new rat's genome against those of its relatives, Lavery confirmed that the large rat was a hitherto unknown species, which he named *Uromys vika* in honour of the rat's local name, vika.

"This project really shows the importance of collaborations with local people," insists Lavery, who first learned about the unknown rat through conversing with Vangunu locals and later confirmed with them that the specimen matched the "vika" they knew.

U. vika are much larger than the black rats that spread throughout the planet with European colonists – the rats found in and around our garbage weigh approximately 200 grams (0.44 pounds), however, the Solomon

Island rats can be more than four times that size, weighing up to a kilogram (2.2 pounds – half the PMW for microcryptids).

Lavery explains, *"Vika's ancestors probably rafted to the island on vegetation, and once they got there, they evolved into this wonderfully new species, nothing like what they came from on the mainland".*

Lavery also emphasised the importance of preserving the rats, not simply for ecological reasons, but for the role they play in the lives of the Vangunu's people.

"These animals are important parts of culture across the Solomon Islands – people have songs about them, and even children's rhymes like our 'This little piggy went to market.'"

U. vika was the first rat discovered from the Solomons in 80 years, and if it hadn't been for Lavery's use of ethnoknown evidence it would have likely been lost to extinction long before being formally recognised by science.

The location where this discovery was made is one of the only areas left with forest that has not yet been touched by logging companies.

"It's really urgent for us to be able to document this rat and find additional support for the Zaira Conservation Area on Vangunu where it lives."

As we have already discussed, cryptozoological animals must be unusual enough, either in terms of their appearance or behaviour, to be noteworthy and therefore suitable for the folkloric process utilised by

cryptozoologists. There must be something out of the ordinary about an unknown animal. A new species of shrew, like any other known, closely related species, isn't going to inspire the kind of awe necessary to make it noteworthy. Or is it?

Enter the Hero Shrew *(Scutisorex somereni)*. This reasonably large mole-like animal resembles a typical large shrew. It has short legs, a slender snout, and small eyes. It has dense, course fur that is grey in colour. It has two types of fur; some hair strands provide sensory functions while other produce scent. The hero shrew aggressively marks its territory, contorting its body to mark objects with its scent. It is thought that the odour repels other members of its species. The chemical it emits can discolour its fur yellow.

The Hero shrew was, until its formal classification in 1910, considered to be an impossible animal. It was long known to the people of the Congo Basin, Africa, who described it as being practically indestructible. It was claimed that an adult person can stamp down on this little shrew with all their might and the little critter would simply scurry away to fight another day.

This still sounds rather unlikely until it's realised that the hero shrew has developed a truly unique physiology. The vertebrae of the hero shrew are thick, corrugated cylinders interlocking on their sides and lower surfaces. The animal's spine has bony projections that mesh to form a very strong, yet flexible backbone. The differences are especially pronounced in the lower back between the rib cage and hips. The hero shrew has 11 lumbar vertebrae, in contrast to a typical mammal which has 5 such vertebrae. The spine of the hero shrew accounts for 4% of its body weight, in contrast to 0.5 – 1.6% for a typical small mammal. The ribs of the shrew are thicker than those of similarly sized mammals and the spinal muscles are significantly different. Its abdominal muscles are reduced, while its spinal muscles are enlarged. As a result, the hero shrew has a peculiar gait with its spine flexing in a snake-like manner.

During an expedition to the Congo region during the 1910s, the native people demonstrated the remarkable strength of the hero shrew to naturalists Herbert Lang and James Chapin. After some mystical preparation, an adult male estimated to weigh 72kg (159 lb) stepped on a shrew and balanced himself on one foot. After several minutes, the man stepped off and the shrew ran away unharmed. The combination of this animal's vertebral strength and its convex curvature behind the shoulder kept its vital organs from being crushed in the demonstration. The feat represented a weight of roughly 1000 times the animal's body weight, the equivalent of a human holding 10 elephants. Relative to body size, the hero shrew's spine is roughly four times more robust than any other vertebrate (excluding its sister species *(S. thori)*, which, incidentally, was only officially discovered in July 2013, from a specimen collected from a local village in the Democratic Republic of the Congo).

Despite its great strength, the hero shrew's spine is easily flexed sagittally (the muscles for doing this are well developed). As a result, the animal can turn 180 degrees within a burrow only slightly wider than the shrew. However, the animal has very little ability to extend its spine or bend it laterally. Its intervertebral joints are five times more resistant to twisting along the axis than a common rat, adjusted for size.

Conclusions

I think we all can agree that cryptozoology is in serious need of some very public and clearly attributable successes. Preferably discoveries made by working cryptozoologists, or by scientists prepared to publicly admit they are clearly using cryptozoological methods. Even with the expulsion of large unknown animals, we still find we have a plethora of cryptid creatures to contemplate and search for, and all the while still utilising the principles originally suggested by Bernard Heuvelmans.

Microcryptozoology doesn't exist. It is for the most part an unwanted aspect of the wider discipline of cryptozoology, and simply represents a real need for change!

By only focusing on the larger unknown animal forms, we are losing ample opportunities to demonstrate that Heuvelmans' methodology is perfectly viable.

I personally believe there are still a few large unknown animals left to discover. Be they yetis, Megacondas, sea serpents, or surviving Tasmanian tigers, compared with creatures of smaller dimensions the numbers left undetected will be minimal. Whereas the chances of discovering or rediscovering small animals is considerably higher.

Writing in *Cryptozoology* in 1984, Bernard Heuvelmans said:

"I have always had to make my books fascinating for the largest possible audience".

Heuvelmans' writing style was purely economic. I believe he selected his cryptids by balancing their biological probabilities with the fantastical elements typically associated with creatures integrated in folklore. The cryptids he selected were exciting, yet still biologically feasible. A zoologist might take seriously some reports of yetis, Bigfoot, thylacines, and Caspian tigers, but we are hard pushed to find any biological scientists prepared to endorse such impossible entities as Mothman, the Dover Demon or the Chupacabras, which are without doubt biologically untenable.

I suggest there is much more to cryptozoology than pseudoscience or monster hunting when the methodology is used in a logical and practical way. It can be optimised to potentially target and subsequently identify new species, which might otherwise go by unnoticed, perhaps even into extinction, or to be discovered later totally by chance (how most new zoological discoveries are in fact made).

Richard Freeman, the Zoological Director of the *Centre for Fortean Zoology*, has this to say about the Mongolian death worm (aka the *Allghoi Khorkhoi*) and how simple it would potentially be for him as an experienced zookeeper, to physically catch hold of a cryptid of this size, compared to the obvious dangers involved in the live capture of large wild animals, unknown or otherwise.

"I think the electrical powers and spitting acid are apocryphal. I think it is a worm lizard or a sand boa, but it is harmless unless you are a mouse. I am experienced in handling reptiles and would catch one bare handed if I saw one and got close enough."

Although admittedly the death worm is usually described slightly too large to be designated as a microcryptid, Richard's rather fitting statement sums up the feasibility and practicality of microcryptozoology perfectly.

I truly hope you enjoyed this preliminary peek into the mysterious and wonderful world of microcryptozoology and hopefully leave with a better understanding of what cryptozoology should be, what it originally represented, and what in future microcryptozoology aims to be.

Texts Cited/Further Reading

Heuvelmans, Bernard. *The Natural History of Hidden Animals*. London: ***Kegan Paul Ltd*.** 2007.

Arment, Chad. *Cryptozoology Science & Speculation*. 1st ed. Landisville, Pa: ***Coachwhip Publications*.** 2004.

May R.M. *How many species inhabit the Earth?* ***Sci Amer,*** 10: 18-24. 1992.

Bebber D.P; Marriott F.H.C; Gaston K.J; Harris S.A; Scotland R.W. *Predicting unknown species numbers using discovery curves.* ***Proc Roy Soc*** B 274: 1651-1658. 2007.

Arnold, Carrie. *New 'astonishing' tarantula has strange horn on its back.* ***Nat Geo***. 2019.

Brown, Barnum. *A luminous spider.* ***Nature.*** Volume 115, page 981. 1925.

Insausti, Teresita C., Defrize, Jeremy; Lazzari, Claudio R., Casas, Jerome. *Visual fields and eye morphology support color vision in a color changing crab spider.* ***Arthropod Structure & Development***, 41 (2): 155-163.

Hastings, J.W. *Biological diversity, chemical mechanisms, and the evolutionary origins of bioluminescent systems.* ***Journal of Molecular Evolution***, 19 (5): 309-321. 1983.

Counsilman, J.J.; Ong, P.P. *Responses of the luminescent land snail Dyakia (Quantula) striata to natural and artificial lights.* ***Journal of Ethology***, 6: 1-8. 1988.

Vrsansky, Peter; Chorvat, Dusan. *Luminescent system of Luchihormetica luckae supported by florescence lifetime imaging.* ***Naturwissenschaften***, 100 (11): 1099-1101. 2013.

Lavery, Tyrone H.; Judge, Hikuna. *A new species of giant rat (Muridae, Uromys) from Vanguna, Solomon Islands.* ***Journal of Mammology***, 98 (6): 1518-1530. 2017.

Hogenboom, Melissa. *New species of super-strong 'Hero Shrew' discovered.* ***BBC***. Retrieved July 24, 2013.

Hatchet, Jani. *Scutisorex somereni*: armoured shrew. ***Animal Diversity Web***. 2013.

Freeman, Richard. Pers comm. 2021.

Unravelling the mystery of an inexplicable Indian reptilian: THE BURU

Saarthak Haldar

In this paper we will be visiting an aesthetic and ravishingly beautiful part of India. There are areas covered by canopies of numerous types of lush green vegetation including many species of trees from different kind of forests, and within those forests lies a world consisting of a mystical land with hills and parts of the Himalayan mountains. The land is embellished with greenery and attracts myriad tourists and visitors toward its forests. A feeling of hidden deep essence of strangeness and undiscovered mysteries can be felt there.

The land described is a part of India named Arunachal Pradesh, a north-eastern state sharing its borders with different countries like Bhutan, Myanmar, and Tibet, and also shares its boundaries with states like Assam and Nagaland. One should know how interesting the name "Arunachal Pradesh" is. The name itself means 'Land of the Rising Sun' or 'Land of the dawn-lit mountains'. When the sun kisses India first, with its warm and soft touch, it shines upon the Arunachal's jungle covered mountains, waking the rural, urban, tribal people and all those known and unknown creatures from their dark, deep slumber.

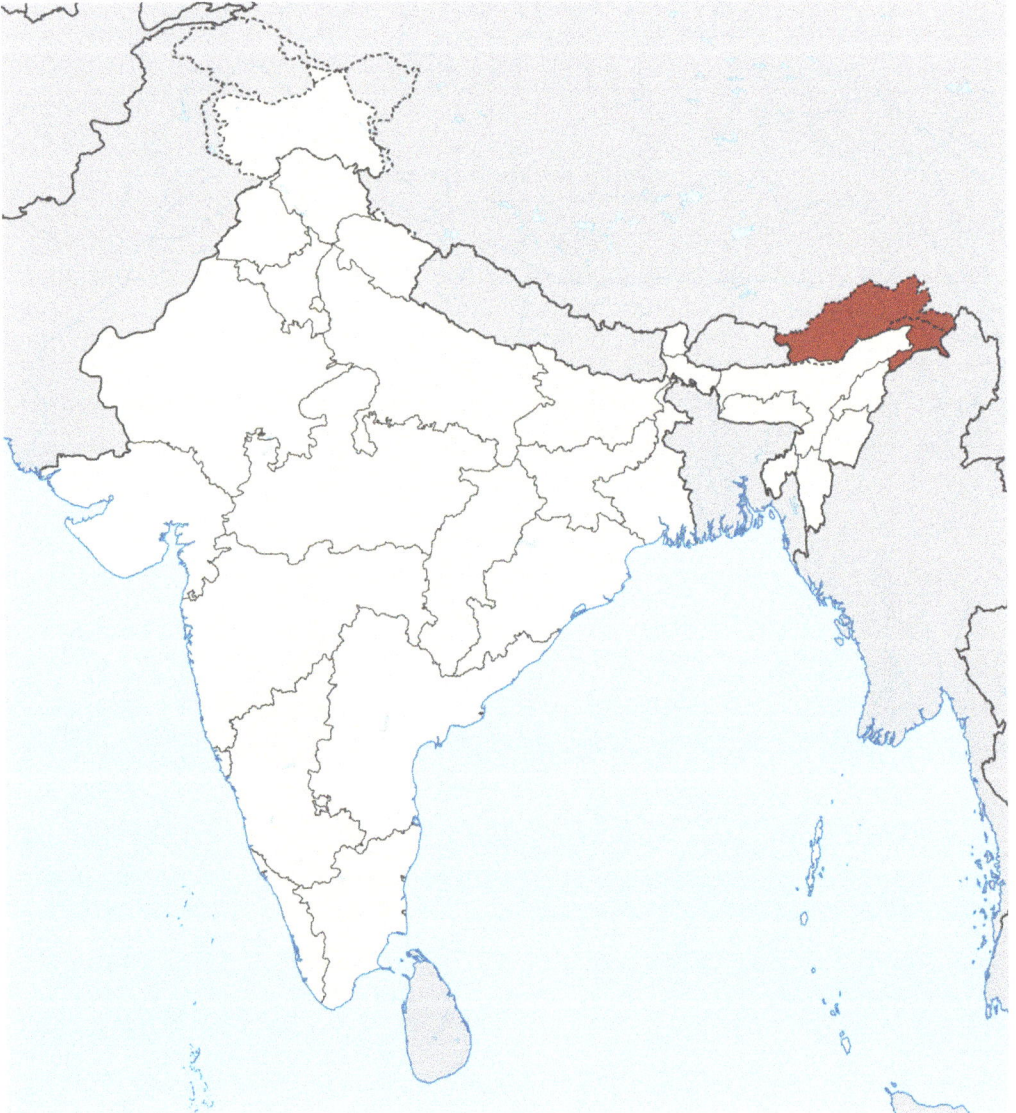

Here resides many diversity of people and hence an example of unity in diversity (Just like the country India itself). Arunachal Pradesh is not just a state full of wilderness, vegetation, mountains and tribal cultures. It's also filled with modern culture, technologies and is going through rapid modernization and urbanisation.

The state has a total estimated population of 1,748,873 in 2021 and it has been estimated that the urban population is 30% of the total. The state is full of modern facilities, institutions, highly-educated people and huge buildings which are expanded and newly constructed daily. But still, approximately 62% of the land is covered in forest, and manages to evoke a sense of thrill inside the heart of every wildlife and nature enthusiast.

43

It is nearly 84000 km^2 in area and the recorded forest area is 51,407 km^2. This area is perfect for exotic, exquisite, exceptional and extraordinary creatures, as it has every kind of ecology and forests imagined.

It stretches from snow-capped mountains to the plains of the Brahmaputra and is a land of lush green forests, deep rivers, valleys and plateaus and even the climate varies from hot, humid and subtropical in the foothills , to a cool, windy climate at lower altitudes and cold and chilly at higher altitudes.

Rainfall is heavy during the monsoon and hence blessed with numerous kind of forests, from hilly forests to tropical forests to temperate forests like, for example:

- East Himalayan Sub-Tropical Wet Hill Forest,
- East Himalayan Wet Temperate Forest,
- East Himalayan Sub-Alpine Birch/Fir Forest,
- Sub-Himalayan Light Alluvial Semi-Evergreen Forest

...and many more different kinds of strange forests, all varying with elevation. There are still so many unexplored and undiscovered forest areas and passes where, there may be unknown species of hidden and unknown creatures lurking and still waiting to be discovered.

But if we talk about *known* creatures then there are also many amazing life forms which have already been already discovered. The wildlife is very rich and varied. Elephants, tigers, leopards, jungle cat, white browed gibbon (in Tirap and Lohit district) and red pandas, musk deer etc (in the higher ranges) are some inhabitants of Arunachal Pradesh. The Mithun (*Bos frontalis*) are found in both wild and domesticated form. It's also home to great Indian Hornbill. Many primates like Macaques, langurs, Gibbons and many kinds of Birds live there happily. There are more than 500 species of fauna known only in Arunachal Pradesh.

If we talk about Ziro Valley. It contains species of leopards like common and clouded leopards, many cats like the marbled cat and leopard cat etc and many different animals like linsang, marten, species of deer and bears and many more creatures. But most of them have low density of population. Grazing animals like deer, pigs and ruminants can be found there easily. Many different species of monitor lizards have also been seen in Arunachal Pradesh.

I was recently reading *The Indian Journal of Tradition Medicine* and found about the recent visit of researchers to a remote valley of Siang district of Arunachal Pradesh, little bit far from the Ziro Valley and found a medicinal approach by Adi tribe using the tail of monitor lizards.

After genetic analysis it was found, that the species used was the Bengal monitor, widely available species in and near Arunachal Pradesh and in Assam. Researchers raised their concerns of wildlife being harmed for unscientific use, and maybe authorities will take some action soon.

The monitor lizards in Arunachal Pradesh are endangered and are being protected now. There are still a few other species of monitor lizards found near Arunachal Pradesh like the Asian monitor lizards (above) Bengal monitor lizards, clouded monitor lizards (opposite), yellow monitor lizards etc.

So now, as you are familiar with the geography, vegetation and animal diversity of the region you will be able to understand the ecology and surroundings better . As you all are aware, how naturally superlative Arunachal Pradesh is, still in these modern times then how it was many years ago and hence the attractiveness of this part of earth attracted some major naturalists, explorers and ethnologists and one of them was Professor Von Furer-Haimendorf. When he arrived here with an excitement inside heart, he bumped into some stories which exhilarated him to the bone. The stories were about a huge, unknown, cryptic reptilian, a lizard like creature residing, hiding and lurking in deep swamps.

A 10-15 feet long or sometimes up to approximately 20 feet, with skin like that of a scaleless fish and has rows of blunt, miniscule spikes running down it's sides and back. Long head elongated into snout. An unknown, unseen, reptilian.

Even before Professor Von Furer-Haimendorf, James Phillip Mills, an ethnographer and Charles Stonor, a naturalist arrived in the Ziro Valley when they heard about Buru and they have even investigated about it by asking the natives. They separately interrogated the residents of Ziro Valley to get proper and correct description of that legendary creature, Buru. They all, including Professor Von Furer-Haimendorf, were intrigued by that creature. A derivation which came from the findings was that the Buru may have spread sparsely in all over Arunachal Pradesh but their prime location is in Ziro Valley, where they have been seen numerous times.

Tucked away in the forbidding mountain scape, an entrancing valley, the Ziro Valley. As those people shared this interesting information about Buru, the info spread like wildfire to other cryptozoologists, biologists, and naturalists of India and across the borders and different parts of the world.

A year after the explorations of Professor Von Furer-Haimendorf, London's *Daily Mail* launched a Buru Expedition funded by the newspaper and Lord Mountbatten

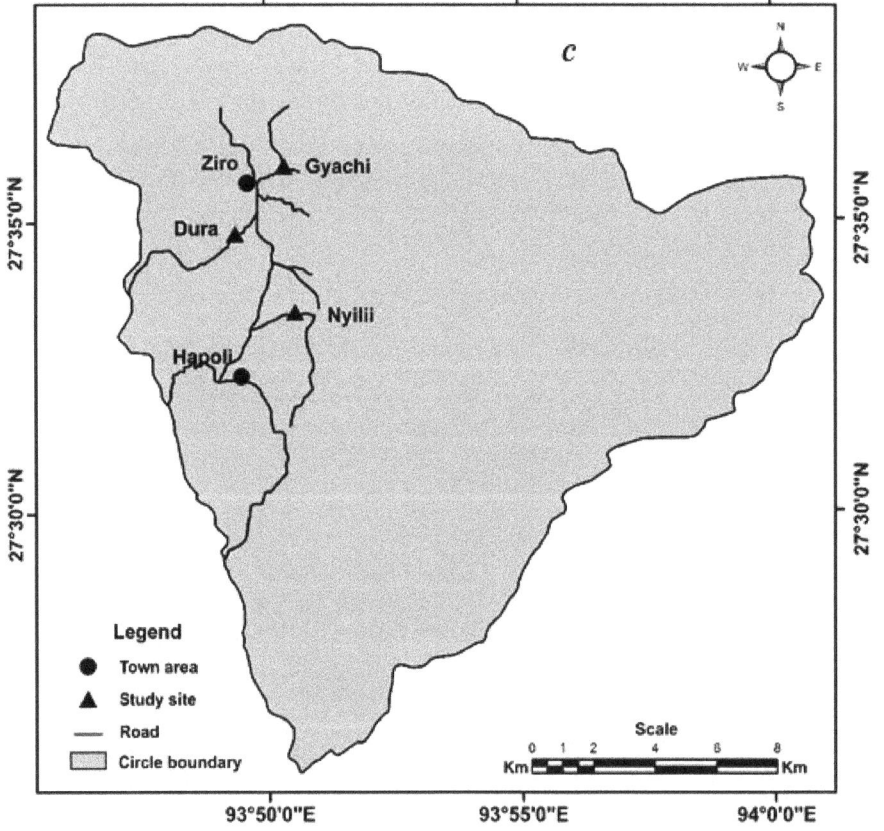

of Burma, who after India's independence when he ceased being the final Viceroy, became the first Governor-General with the Hope of finding some groundbreaking evidence on Buru's existence. The expedition consisted of notable researcher and zoologist Charles Stonor accompanied with a notable journalist Ralph Izzard, who later penned down the entire expedition on Buru which was published in 1951.

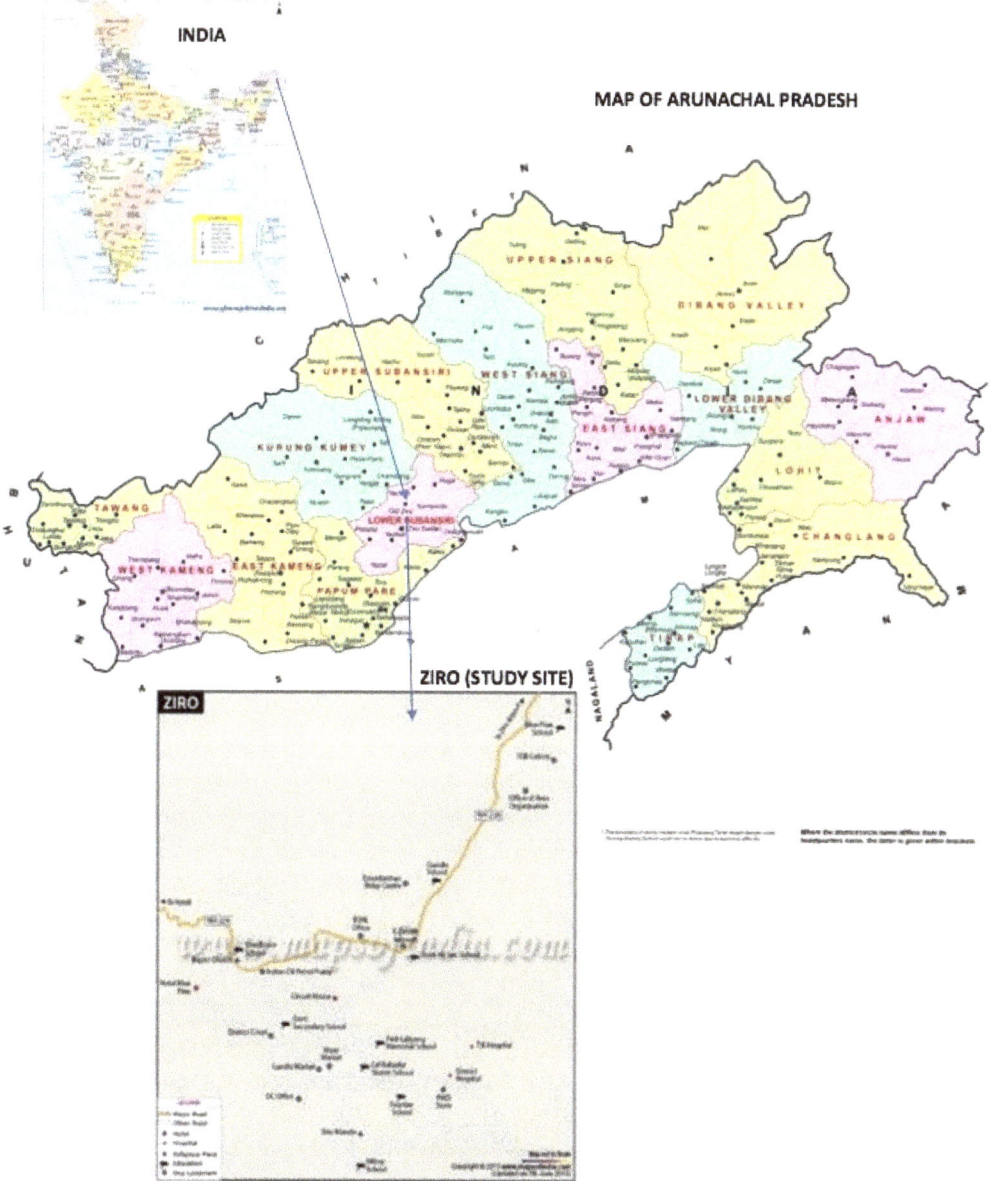

INDIA

MAP OF ARUNACHAL PRADESH

ZIRO (STUDY SITE)

ZIRO

This all actually started when Ralph Izzard was sharing his findings with A.P.F Hamilton of the Indian Forestry Service and then he told Ralph about Stonor and his discoveries in the lost valley. Ralph then contacted Stonor by letter and as letters were exchanged between the two, their relation and thoughts about the existence of Buru both strengthened.

Even after all these, the team was not able to bring back any solid evidence but still found some testimonies of earlier encounters with Buru. They also found some conflicting stories from the Nishy People, one of the largest ethnic groups in Arunachal Pradesh. Perhaps, it can be possible that both groups, Apa Tanis and Nishy were confused about the Buru's features due to fewer encounters, or perhaps both were talking about two very different creatures and hence their opinions conflicted. Renowned cryptozoologist Bernard Heuvelmans was convinced at last even by these evidences and concluded that these unidentified creatures may have only *recently* became extinct. However many still think that Buru is still lurking and slithering somewhere in the swamps and grasses in Arunachal Pradesh.

While Ralph Izzard, from the *Daily Mail* expedition, separately questioned the leading witnesses and recording the data in detail, he discovered more facts about the Buru.

The teeth were basically flat but a pair of teeth from the upper and lower jaw were large and pointed. A mottled bluish-blackish colour with an elongated neck. Black and dark blue in colour with white spots on the body and the underbelly having a whitish hue and shade. With four short legs which were highly clawed, it looked terrifying.

It also possess a lengthy and powerful tail. The texture seemed like having rows of armoured plates. The creature lives partially in water and can make haunting loud, hoarse, bellowing sound and waved their heads and necks when they do so. Though they were fierce and dragon-like, they tended to keep in themselves. They are not generally very aggressive, however they were framed as the culprit of some attacks on humans. They mostly hid themselves in the swamps and remained in the mud when the swamps dried up.

During rain they come out of their hiding places to play, prey and perhaps show happiness. Many people described it as a huge lizard-like reptile with a forked tongue. In some accounts it raised it's head out of water occasionally and many witnesses claimed that it basks on the bank of the river. The witnesses, however, said that they hadn't seen it feasting on fishes but that it ate other aquatic creatures. Young ones were reportedly born alive in water. A few people said that although they are sluggish, they are capable of manoeuvring at short bursts of speed.

It has been said that Buru are not naturally aggressive and that they like to keep to themselves, but one day a hunter threatened the young ones of a particular Buru and that then the Buru's maternal or perhaps paternal instincts kicked in and it retaliated by attacking the hunter with its powerful tail and destabilized and hence drowned him.

All these facts gathered after questioning convinced them to confirm the existence of the mysterious creature. There are many assumptions given by biologists and cryptozoologists who had been investigating, or who had read about the Buru. Tim Dinsdale, a British cryptozoologist said that in his opinion the Buru was some form of giant crocodile, some unknown species of crocodile by the Apatani People. Both Bernard Heuvelmans and Roy Mackal regarded Buru as a large Komodo dragon-like monitor lizard. To support this theory they said that fossils of such creature have been found in Indian subcontinent. Heuvelmans noted that similar reports of creatures like the Buru also came from Burma and they might relate to a reported lizard-like monster in the Mekong river.

Heuvelmans also noted that similar creatures named 'Jhoors ' were reported in western India where they seemed to merge into the Iranian traditional dragon-like

creature, Azi Dahaka. Author George Eberhart noted rumours of a similar creature in Tigris Marshes of Iraq called the 'Afa '. Many more different zoologists and

cryptozoologists who read about the Buru came up with different conclusions , such as Ralph Izzard who suggested that it was a surviving dinosaur of some type . Others suggested that it was unknown species of Bony tongue fish similar to the pirarucu (*Arapaima gigas*).

Some even speculated that it's a type of stegosaurus, such as Tuojiangosaurus (above) or Wuerhosaurus, (below) on the basis of spikes and plates on its back. Another one of the famous, gifted and one of my favourite zoologists and cryptozoologists, my friend Dr Karl Shuker suggested that the Buru was a giant

lungfish and goes on to say that this provides more comparable match, not only in terms of body structure but also with regard to the behaviour.

Only this explains the Buru's alleged ability to survive hidden at bottom of lakes during the dry season and keep submerged in mud and perhaps, it's bellowing might be caused by its ventilation of air.

All these theories are astounding and provides insights on the Buru's features. When I started digging deeper and researching on Buru's reality, I came upon something similar to one of the theories above. I was persuaded by that one theory which claimed that Buru is kind of reptile similar to monitor lizard or perhaps, as I started to believe, it *is* a monitor lizard. There is a 'huge misunderstanding' which also, we will touch upon later.

There are still few species of monitor lizards which inhabit in and near Arunachal Pradesh for example, the Asian monitor (*Varanus salvator*) with dark brown or blackish color sprinkled with yellow or whitish spots, the Bengal monitor (*Varanus bengalensis*) with series of crossbars or spots on neck and back, the clouded monitor (*V. nebulosus*) with yellowish spots on Dark greyish base and yellow monitor (*V. flavescens*) with yellowish brown and irregular dark markings.

Now I would like to discuss why I believe Buru were actually a monitor lizard. I came into this conclusion after researching and reading the tribal scriptures and other written pieces of researchers, I also contacted a native and discussed about Buru. So, as the tribal people described Buru's length to the researchers, their statements varied greatly. There is no perfect consensus regarding Buru's length so it's a bit difficult to form a solid conclusion but if we establish that the Buru's length is not much less than 10 feet as most of the Apatani people claimed that it is approximately 10-15 feet in length. In India and near Arunachal Pradesh, after the giant crocodiles which can be 15-20 feet (sometimes more than 20 feet) in size, the next largest is the Asian monitor lizard with can attain a size of 8-11 feet and there is no other reptilian of that size. The largest specimen of Asian water monitor lizard on record from Sri Lanka measured more than 3.20 meters.

Monitor lizards of any species can grow incredibly large in comparison to most lizards. When a species becomes isolated in a habitat with very few other big predators, they tend to become much larger. As big predators are few, the ecological niches are left a bit vacant and then it becomes available to small animals and with nearly negligible limiting factor. They become larger and big size becomes a common trait in the population. This has been seen mostly on isolated islands but it can happen in any region with fewer or absent big mammalian predators. Everyone will have heard about "Natural Selection" in which animals better suited to certain environment are selected by nature and they thrive there nicely. Natural selection occurs and the bigger lizards which have an advantage are selected. The traits of larger body can be due to larger gut size and larger gut is required to process the food of lower quality which is found in those areas and also due to this trait they get a better metabolism and when they get better quality food, they are able to absorb it nicely and get better nutrition than others and hence they grow large .

Another point is that the heat and warmth helps the ectotherms like the monitor lizards to grow bigger than normal size and also increases their length of growing period by activating their body enzymes to the highest potential.

So, the climatic conditions favourable for monitor lizards or any other reptilians like them in which they can easily flourish and grow is one with extreme warmth and humidity. Every monitor like the Asian monitor likes to bask in temperature between 70-100 degree Fahrenheit with the ambient temperature lower than the 70-80 degrees Fahrenheit and with near 60-70% humidity. These conditions would help and influence the monitor lizards to grow nicely and to reach and sometimes surpass their regular limits. We can see the climatic conditions of Arunachal Pradesh would be more than perfect for such creatures. Here the climate is highly hot and highly humid at lower altitudes and valleys such as the Ziro Valley, where apparently the Buru lived happily, are covered by swampy dense forest and the humidity and rainfall is amongst the heaviest in the country. It's like heaven for that kind of reptilie and as its been said, it was heaven for the Buru too.

These are just *some* suggestions as to why Buru could be monitor lizards and how the monitor lizards in specific areas can grow bigger than other members of its own type. There are only a few creatures which can fit the description of Buru and they are mostly reptilian creatures and few fishes and as far as I know, no fish or fossil of fish has been found which ever lived in the Northeastern part of the Indian subcontinent with characters similar to Buru that was so huge, except one, about which we will discuss a bit later.

There are many fishes with a huge size similar to that of the Buru, but they don't fit into the description. So it's a possibility that the Apatani people were talking about a monitor lizard.

It has been said by Apatani people that the Buru possessed an ability to survive by hiding in the mud which was near the bottom of the lakes, when the lake partially or sometimes mostly dried in the hot summer seasons. I have witnessed that when the season becomes dry; the water in the lakes evaporates, leaving behind the mud and as alleged, the Buru, with the capability to hide itself under the muddy surface could easily survive.

And while talking to a few crocodile experts who were experienced in fieldwork, I was informed that crocodiles adopt two ways to survive the heat and drying of water bodies. Most of them bury themselves in the mud at the bottom of the pond or lake to retain their skin moisture and to protect themselves from the scorching and burning sun but some of them try to escape the drying water body and try to walk to the nearest water source in the area. And that they usually travel at night when it's cooler and when no-one can see.

So I believe that by focusing on and associating a few facts;

- Firstly, that crocodiles bury themselves in remaining mud and water as the Buru is said to do.
- Secondly, as crocodiles travel at night the Apatani people would not have seen them leaving the water body and would have come to the conclusion that they are still there hiding in deep mud as they do in the daytime and hence come to the conclusion that they survive by hiding inside the muddy soil.
- Thirdly that Apatani people misunderstood the difference between crocodiles and monitor lizards and thought that they were the same and

never differentiated between them. Therefore, they called both crocodiles and monitor lizards 'Buru'. So we can derive the conclusion of what the Buru is according to the Apatani people, and this adds to our understanding of the true reality of the Buru.

There is also the possibility of a big confusion between Buru and crocodilians. I will tell something associated with it later when the "huge misunderstanding" will be described thoroughly.

As the tribal people have said, Buru have rows of small, miniscule spikes running down its sides and back. It could be an illusion or a misinterpretation to be more exact. Monitor lizards have big and major skin folds on its sides and if we talk about the tail, the tail is very flattened laterally. As the Apatani people have mostly seen Buru in the mud or swamp, so while partially submerged, it could give illusion of some special protrusions and perhaps, while smeared and pasted with mud, the illusion must have amplified and as Apatani people never inspected it closely, they started thinking of them as mini spikes and formed an idea based on that illusion.

Actually as I know, the Apatani people tended to remain in their own areas and were scared of the Buru and perhaps the Burus were afraid of them as well. So the clear observation and inspection has never been done.

When I was observing a monitor lizard, I found, no matter how pointed the teeth are, while seeing it feels like they are mostly hidden, it's very hard to see them completely and clearly inside the mouth from any distance. Their teeth are not as visible as those of a crocodile. Amazingly there is a similarity between this and the statement of the Apatani people when they said that Buru have mostly flat teeth, and are not big, pointed and protruding. If we totally believe the Apatani people's claims, and if we believe that the Apatani people inspected the teeth of Buru by studying their dentition closely, which is perhaps very unlikely, even after that the possibility of Buru being a monitor lizard can't be denied totally.

Some adult monitor lizards (Varanus sp.) predate on snails, clams and crustaceans. To crush their well-packed food, some species develop broad, flat teeth. Eggs are also one of the main food of monitor lizards; the flat teeth may help them to grasp and crush these eggs as well. These happen because in many species of monitor lizard, diet changes with age, and this is reflected by changes in dentition. Very young monitors have long, sharp teeth housed in a relatively small skull. As the animals attain maturity the bones of the skull become much thicker and the sharp teeth give way to the broad, crushing teeth typical of the adults. The front teeth remains pointed but the teeth in the back turns into more flat and levelled for grinding. This feature has been extensively studied in Nile monitors but many different species have shown this feature and in fact any individual or small population of any species can show this attribute, due to transition in feeding niche and comestibles, hence causing a polymorphism (a genetic variation resulting in the occurrence of several different forms or types of individuals or groups among the members of a single species) in population.

All these conditions above could have transpired causing flat dentition of monitor lizards but I prefer the first scenario in which I believe, due to unclear vision and perspective, they became confused and formed a notion that Buru have flat teeth.

Some tribal people have made claims about the Buru's tail and how it uses it, and says that there is a noticeable similarly with monitor lizards. As the Apatani people said, Buru had long, lengthy and powerful tails and used it mostly to deter it's predators. When a hunter threatened it's young ones, it lashed at the man and drowned him by destabilizing, by forcing him to lose his balance with its powerful tail. All of this description matches with the monitor lizards and the behaviour matches with them too. I recently saw a riveting and enthralling video in which a leopard was trying to prey on a relatively small monitor lizard, perhaps a young specimen, and the lizard was trying its best to defend itself by turning its back towards the leopard and was relentlessly trying to whip the big cat with its tail.

At last the lizard lost the battle for life; the big cat grabbed it by its neck and left. It was a really poignant moment but that's how the food chain and food web works. Nonetheless, it was a good example of how monitor lizards use their tail and it's exactly how the Apatani people described the Buru's tail. Both whip their long, puissant tail to fend off and ward off their potential threats. However crocodiles do the same often when they feel that their young ones are threatened.

One statement of the Apatani people claimed that, despite their fierce and terrifying dragon-like appearance, they tended to keep themselves to themselves and were generally not very aggressive and attacked only when provoked and threatened and if we analogize the resemblance and similitude with monitor lizards we will be able to cognize that monitor lizards too depicts similar behavior.

They are not very aggressive, they are naturally shy and like to stay away from humans and they generally retreats and run away from people as I have witnessed in many cases. But they can deliver very nasty bites or lashes to anyone or anything which threaten or provoke them.

If we talk about fishes, for example, in case of lungfish it is written that, they are not particularly aggressive as well, despite not really being aggressive, they will eat nearly anything they can get into their mouths. Big mouths and huge appetites characterize them. They chase any moving thing which can be there prey. They can even bite the chunks out of creatures bigger than them. They mostly rely on their gnashers for attack and defence and not on their tails normally. Many, who raise them, say that they are unpredictable with their attacks.

There are few more assertions about the Buru like the one which claimed that during rain they come out of their hidings, just like the other reptiles like the monitor lizards, who come out of their hiding places during rain due to many reasons.

- Firstly, for mating, as in India, Bengal monitor lizards and Asian monitor lizards come out to mate in June, July and August and according to climate of India extreme monsoon and raining starts from June to early September.
- Secondly, for prey, there are plenty of prey which come out during rain or rainy season
- Thirdly, due to rain the burrows of those creatures gets flooded and it forces the creatures out of their burrows.

According to the Apatani people, Buru's young ones are born alive. There is a possibility that due to the lack of proper knowledge about the Buru they had become totally confused. Monitor lizards are oviparous: they lay many eggs at a time, but they cover their eggs with soil or protect it by hiding it in a hollow tree trunks and stumps and when the time arrives, the eggs hatch and babies come out. Only the few fortunate, or the very persistent observers who are there at that very moment, have the blissful opportunity to see the laying of, or the hatching of the eggs. Maybe due to Apatani's lack of continuous encounter with the Buru or lack of proper knowledge of the reality of the Buru they have assumed that babies are born alive.

One statement claimed that the Buru does not eat fishes but perhaps takes other small creatures or maybe vegetation and that also it's what monitor lizards do as well. Unlike fishes who's diet mostly consist of other small fish, the monitor lizards can also prey majorly on numerous small creatures like lizards, frogs, small snakes, rats, eggs etc and some even eat fruits and other vegetation, depending on where they live.

There are many other points describing the characteristics of Buru, told by the Apatani people and written about by researchers like the long head elongated into snout, elongated neck, mottled bluish-blackish white, black and blue coloured with white spots, underbelly having whitish shade, having short forelegs with big claws, a lengthy, powerful, long tail, and forked tongue. It's nature of basking in the sun on the banks of rivers, capability to manoeuvre at short bursts of speed, each and every point fits perfectly and impeccably in place just like some lost puzzle pieces. Every point written above seems like it's finely describing the monitor lizard without any flaws. These points nearly clears the blurred and misty reality of Buru.

All these facts described in this entire article are already written and can be found in any link of encyclopaedias on Buru or papers published by the prior researchers . Very few facts can be bit different or new because I have written this also after contacting the natives of the Ziro valley and a person named Paiang Tage, helped me to find and confirm the chronicled facts by asking the elders. It's true that most residents don't know about the Buru but a few native elders retained the knowledge of this creature.

Now, as I have mentioned earlier in this article about a "big misunderstanding " so now I think it's the time to get to that. As I have already told you, the Apatani people were confused about the Buru and misunderstood some of its characteristics due to lack of proper observation, and perhaps even out of fear. But there is one more thing.

James Phillip Mills and Charles Stonor and even Professor Von Furer-Haimendorf investigated the Buru by asking the natives separately, in isolation, that's where misunderstanding started. The Apatani people, during those days, never knew how to differentiate between crocodiles, like Gharials (Fish eating crocodiles), and monitor lizards. They referred all of them as Buru. For many years no crocodiles have been detected in the Ziro valley because they are endangered and human encroachment restricted the crocodiles. A few people claimed that there were no crocodiles found in the Ziro valley at the time when the Apatanis arrived, but it's not necessary, as there were many rivers near Ziro valley in Arunachal Pradesh and it is possible that crocodiles arrived from the big river Brahmaputra, where crocodile (Gharials) populations were high. They must have arrived into small rivers near the Ziro valley and then eventually they arrived at Ziro Valley, and took shelter in the marshes and swamps which stored enough water which would have provided a nice shelter to those crocodiles. Those crocodiles (Gharials) could also have travelled between the rivers and swamps as the crocodiles have been found travelling few kilometres on land.

When those researchers asked the natives separately, they were discombobulated and described everything they knew about the Buru. Some described the Buru by providing the characteristics of crocodiles (Gharials), some described features of monitor lizards and some amalgamated the characters of monitor lizards and crocodiles into a single creature.

When all researchers combined the cumulated info, the birth of a strange organism took place. I believe, this may have happened in the Ziro valley. This revelation can shed light on few features of Buru, like its tendency to stay mostly in mud, small spines on its back and sides, its texture which feels like armoured plates, it's loud bellowing (we all know crocodiles can bellow loudly). Crocodiles (Gharials) protect their young at any cost, they also slap the creatures, with its very powerful tail, which comes near their young, so it can be also possible that it was a crocodile who drowned the hunter. This also explains why few tribals described that Buru has length of approximately 20 feet.

Still, it's evident that the Apatani people were mostly trying to describe monitor lizards, as most of the points about the attributes of the Buru collides with the features and qualities of monitor lizards like its black-blue shade with white yellow spots, forked tongue, elongated neck, clawed feet etc. We came across many theories which tried to deduce what the Buru can be and there are some theories with which I agree too like the theory proposed by Heuvelmans and Ray Mackal who regarded Buru as big monitor lizard, and the theory by Tim Dinsdale who regarded the Buru as some kind of crocodile, we can also call it true to some extent, perhaps.

Heuvelmans claimed that the Buru or Buru-like creature is related to a monster in Mekong river but as far as I know, the monster is more elongated, long and flexible. It was more like a serpent or maybe some kind of elongate fish. Jeremy Wade claimed that the monster is a huge catfish, which could be possible because it's somewhere related a bit to the information of the ancient explanations. Heuvelmans also noted that similar creatures were found and reported in western India about which we have prior discussed in brief. The "Azi Dahaka " is a creature from Zoroastrian Persian mythology. It has been argued by Heuvelmans that Azi Dahaka is similar to the Buru but as it's been said by the people who are aware of this mythological creature, it's more like a serpentine Chinese dragon. Azi Dahaka, as the name suggests, Azi comes from Ahi (Sanskrit language) which means serpent and Dahaka comes from Persian meaning Half serpent king and it feeds on humans which Buru doesn't. So, now we have few reasons to believe that those creatures are not fit to be compared with Buru. Even though, Heuvelmans was perhaps almost correct with his assumption that Buru can be a monitor lizard.

There is one more creature known as "Afa" from the Tigris marshes, popularised by an author. As there is no deep knowledge available about Afa but as I know it shares many features with Buru or perhaps the monitor lizards, so it can be possible that Afa is kind of monitor lizards as Iraq has some species of these lizards. Ralph Izzard speculated that the Buru is a surviving dinosaur of some type. Some other researcher have said something similar, that it can be type of stegosaurus such as a Tuojiangosaurus or Wuerhosaurus on the basis of spikes or plates on the back. I will not try to deny that ancient fossils of stegosaurian dinosaurs has been found in India and China. But no new fossils of that creature have been found near Assam or Arunachal Pradesh and it's next to impossible that any dinosaur have survived for that long time by creating offspring. And as the tribal statements suggest, the Buru had small spines so it's can't be *any* species of stegosaurus which had such huge spines and plates on its back. The Buru also had short legs compared to Stegosaurian dinosaurs.

One more theory was that Buru is a kind of lungfish. This is a theory which I love and am enthralled and entranced by. I even believed this before digging deeper into the subject. Let me tell you that whenever I propose any idea I try to prove myself and my idea wrong but when I get defeated, then I can start believing with my heart on that idea. When I was studying more, with my new fascination about the fascinating and fantastic creature, the lungfish, I found a few things about lung fish which ignited the hope that it can be the Buru. Eventually, more points and features were in the favour of monitor lizards. But there are a few features of lungfish that really made me enthusiastic, and are also bit similar to the Buru. As already pointed out by Dr. Karl Shuker that the intake of air readily perceived visually by the movements of mouth and throat can also be audible. Another point was that, as it's been said that Burus are not totally piscivorous, and the lung fish also rely on diet such as insect larvae when juvenile and after growing to 2ft, snails becomes its main diet.

He also pointed few physical similarities like the spines on the Buru, which can be actually a fin of lungfish mistaken as spine and the behaviour of lungfish to stay inside the swamp for few amount of time which is identical to the Buru. And few more points. As I said I was trying too, to prove that the Buru is a lungfish. Whilst doing this I added few more points. If we see the climatic conditions, like temperature and humidity of Arunachal Pradesh, it certainly favours the lungfish but also the monitor lizards. The temperature in the monsoon in the lower regions and is approximately 25-35° Celsius and sometimes higher. In summer it reaches up-to 40° Celsius. As the research done by two scientists, Ashley Seifest and Lauren Chapman on lungfish showed, the SMR (standard metabolic rate) was very high at a temperature of 30° Celsius and bit above. As the suitable temperature activate their enzymes and their metabolism to the fullest and fishes eat more often as the digestive system functions quickly and the fish can become larger than its normal size, it can also happen with monitor lizards, because of enhanced metabolism due to the temperature, they can also become huge.

The largest lungfish found in Australia in records was nearly five feet (approximately) in length and if we talk about African lungfish, the yellow marbled Ethiopian lungfish species has been known to grow around two metres (approximately seven feet) in length. Not small but not as big as the Buru. If we see another instance, researchers James Kirkland and Kenshu Shimada estimated, a new Ceratodus (genus of extinct lung fish) was approximately 4 metres long which now holds the title of worlds largest lungfish and dates between 100 million to 160 million years ago. A resident of central Nebraska found the teeth in 1940 and

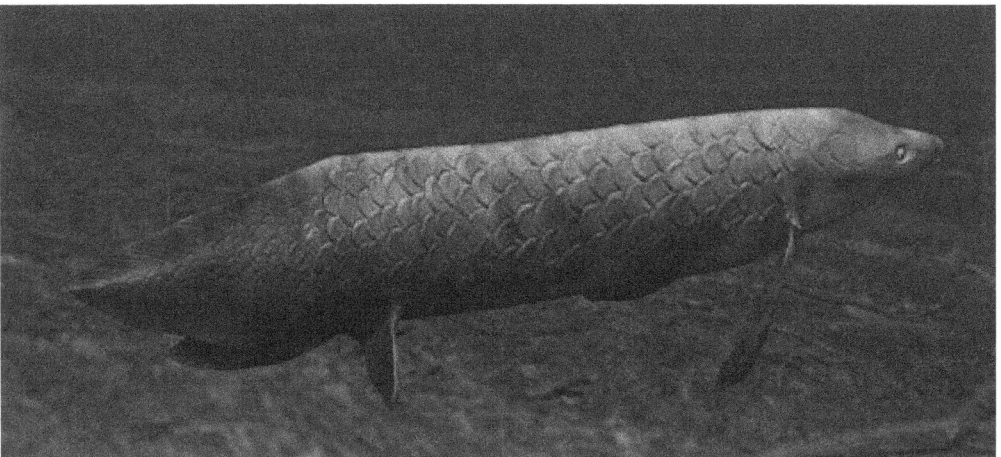

submitted it to the museum. Since the fossil is not from the right geological area so the giant lung fish may have lived hundreds of kilometres away, in Wyoming.

Maybe it was large but it went extinct millions of years ago and it lived in the United States, approximately 8,431 miles away from India. In my opinion, existence of a similar creature so far away in Arunachal Pradesh, without leaving any fossil trail, is quite unlikely. Even If we try to consider the possibility of mutation in the pituitary gland of any fish and hence change in amount of GH (growth hormone) and making it large, more spiny, altered aggressive behaviour etc, (which can also happen with monitor lizards) it's highly unlikely. There is a less than negligible chance. And if it had happened, it would have affected only few or single individuals. I stumbled upon one more fact while disproving the monitor lizard theory. When its been said that Buru could be a Lung fish, many opposed that by saying that there are no lungfish found in India but during Carnian (upper Triassic period) a few species of lungfishes existed here, after which, no lungfishes ever existed in the Indian subcontinent. Lungfishes now only exist in Africa, Australia and America. The lungfishes (Dipnoi) which made India home have only left their remnants and vestiges. Numerous tooth plates of different types were recovered from upper Triassic Tiki formation of India. Sharp crested tooth plates are assigned to a new species of Dipnoan genus Ptychoceratodus. Another tooth plate with different type of ridges has been identified as a type of Gnathorhiza. This is the first record of Gnathorhizid fishes from the Triassic sediments around the world. The Tiki aquatic realm was inhabited by different types of fishes including the omnivores and carnivorous Dipnoans and other bony fishes. The palaeolatitude in the southern hemisphere, was where several genera co-existed in India. But in Holocene epoch of Quaternary period, in the Cenozoic era, in which we are living now, we haven't found any lungfishes (small or giant) in India fossil history, even after so many intense investigations, explorations and research. When I was studying the lung fishes, I realized, the theory about monitor lizard being the Buru was more persuasive and I felt that it is more weighty.

As it's been said that the lungfish is more comparable in terms of body structure but if we see the claims by the Apatani people, we will find most of the body structures and features different and dissimilar like the colour, it's elongated neck, four legs, heavy claws, forked tongue, it's nature to fight with its tail, basking in the sun etc. One statement was that Buru give birth to its young ones which are born alive in water. As we all know, most fishes, even lungfishes, lay eggs in a nest and male guards it's eggs and it's young ones.

Eggs are most abundant during spring seasons. Perhaps, spotting the eggs will not be particularly difficult for the Apatani people but it can also be possible that due to aquatic plants and mud, the Apatanis were unable to see those eggs. There is only a slight chance of this, but I suppose that it is still possible. Also, the Apatanis were prodigious fisherman and fish farmers, fish was their one of the staple diets and they were very knowledgeable about different kinds of fishes. As humans, all have pattern recognition and inductive reasoning and thinking abilities, these abilities are speciality of the human brain, to not only find and identify patterns but to figure out in a logical way what those patterns suggest about and what will happen next. So the Apatani people were fisherman with proficient pattern recognition skills in perceiving and understanding the nature including fishes. They caught and cultivated fishes for generations and so, if we consider the lungfish, they have powerful, elongated, and snake-like, or eel-like bodies, so it's less likely that they will be deceived by fish-like creature and will misunderstand it as some lizard like reptile. Notwithstanding, I am so astonished and awestruck by the theory of lungfish that I understand why it has been given utmost preference and priority after the crocodile and monitor lizard theory .

So, actually, with the scientific development of India, Arunachal Pradesh also developed and flourished scientifically and educationally. That state has also given us many biologists and ecologists. It has many research institutions also and hence researchers assiduously try to discover new species or remnants of any ancient and prehistoric species. So, it's bit doubtful that now in this era, any evidence of huge and abundant creatures like the Buru which once lived in the Ziro valley which is now a populated and educated area, will remain hidden and unknown, I am not saying that it is absolutely impossible that any evidence can stay hidden, but it is very less probable. Now, if we shift our focus to monitor lizards, we will find something rather interesting. As the claims of the Apatani people suggests, the Buru became lost from the Ziro valley, after their arrival And if we see the conditions of monitor lizards, they became endangered in the whole of the Arunachal Pradesh after the arrival of human population and the increase in human population there. I believe, as the Apatani people started cutting the trees to do cultivation they unknowingly started damaging the monitor lizard population. As young lizards are vulnerable and seek shelter dwelling in between the trees, hollow trees or tree caves to hide and avoid predators and the cannibalistic adults. Due to clearing of trees, the number of young monitors declined. Now, as the Apatanis confess themselves they drained the marshes and threw stones at them to eliminate them all. So, these all add up and made the population of monitor lizards extinct from that specific place, the Ziro valley. Perhaps the immigration of monitor lizards also decreased for some reason, like the distance (if there were any monitor lizard in Arunachal Pradesh or Assam, they were so far that there was no way they could travel and repopulate the Ziro valley) and disturbance and due to this the Apatani people started seeing very few Buru or no Buru at all. When I was writing of the Apatani people's confession of how they tried to eliminate the Buru. Their statements really saddened me.

The Apatanis said that they were getting too many attacks and confrontations from the Buru. As it was not the Buru's fault the Apatani people decided to take drastic measures. They built a series of trenches below the water level, allowing gravity to do the work of propelling water down and hence out of the swamp and then filled the swamps with dry soil and rocks, by this way they drained and dried out the marshes and swamps. When the Burus were not able to hide themselves, they were cornered by the Apatani people who surrounded them and stoned them to death. If we imagine the massacre it was carnage. It was really heart wrenching to know that it had been home to Burus and they had lived there happily for generations but then the Apatanis arrived, threatened them and their young ones unknowingly and as a basic animal behavior and instinct they attacked the people, and when they attacked the Apatanis slaughtered them into extinction. Perhaps, it's what Apatanis believe; they assumed that Burus went extinct but their illusion broke as a Buru was sighted again by a young woman who saw it one night near the spring, when she was fetching water from the spring and then she informed her father and next day whole village arrived and filled the spring with stones and clay. Maybe few or many Burus escaped that mass-slaughter and hid themselves in the burrows and came out when the Apatani people were gone and perhaps evacuated the valley. Their own home valley became their Death Valley and they were forced to forage furtively and secretively and even evacuate their own home.

Still, the probability is the Burus are monitor lizards (and could have been crocodiles (gharial) too) and after the stories of massacre, it is very pleasing that monitors and gharials are still roaming this earth, but if there is a minute percentage of a chance that Burus are kind of some new species then we should pray to God that they should not go extinct and intense expeditions and explorations should be carried out in order to find them and conserve them as their existence will push the limits of zoology and biodiversity beyond and will provide humans with new scientific knowledge and understanding.

Exploring the Prospect of an Unidentified Species of Reptile within Navajo and Hopi Lands:

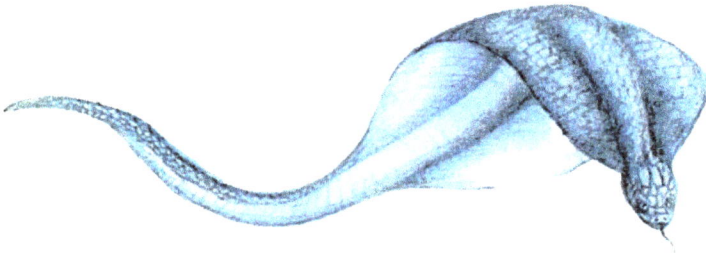

In Search of

Tł'iish Naat'Agii (Snake-That-Flies)

By Nick Sucik
Edited by Russell Bates
Images by Aleksandar S. Trivich

September 2003 (Revised)

Since early in the Spring of 2003 I have been conducting research into what appears to be a form of unclassified and unidentified reptile present within Navajo and Hopi lands in Arizona. In presenting a general report from my ongoing investigation I hope to:

a) create awareness as to the presence of the potential species in question
b) receive co-operative assistance and/or a formal endorsement toward current efforts to locate a live specimen, and
c) understand the Tribe's position toward seeking out potentially unclassified animals within the Hopi Reservation.

INTRODUCTION

As of March 2003 I have been employed as a member of the Peace Keeper project assigned to survey and to assist with the needs of Navajo families living within the Hopi Partitioned Lands. Being originally from Minnesota myself and having been stationed in Hawaii with the US Marine Corps, I found it extraordinarily easy to take great interest in the comparatively unique landscape and ecology of northern Arizona.

During the course of my stay in this region, word has come to me from both the HPL and the Navajo Partitioned Lands residents regarding rumors of strange, seemingly out-of-place animals allegedly sighted from time to time in certain areas. Considering that these vast stretches of terrain relatively are uninhabited, it would not seem impossible that there yet may exist fauna undocumented by modern science. Hence, I've made careful inquiry among the locals to see if there wasn't more to these rumors, that is, beyond mere loose gossip. I in turn have come to hear a long and rather colorful assortment of strange creatures allegedly seen or at least spoken about. This anecdotal bestiary includes within it snakes the size of telephone poles, scorpions the size of dogs, monitor lizards lengthy enough to span across a dirt road, nocturnal gorillas, stalking hyenas, skulking leopards, howling wild men, hostile mermaids, lurking water serpents, and elusive miniature horses that only are seen during early summer mornings.

Some of these alleged entities might be only the result of misidentifications, coupled with local excitement (if not hysteria) and generalized exaggeration.

Case in point: during June 2003, I heard rumors of a hyena being seen by locals in the Big Mountain area. The beast was accused of devouring newborn calves and colts alike, along with having taken down a few sheep as well and, most disturbingly, on one occasion being observed eating dog carcasses. A local man was said to have tracked and killed the beast, the carcass of which area residents then came to view. Some said it was in every way like a hyena, while others said it more was like a big cat. I reasoned that, if there were any question left lingering over the identity of the animal, its bones should be recovered and brought in for examination. Such efforts were proven unnecessary for, when I spoke to the man in question, he promptly identified the animal as a dog. It was a very unusual dog both in appearance and behavior, but it still simply was a dog.

In spite of this sobering example, there are some stories of particular creatures that have such a consistent pattern that one wonders if there possibly is factual basis behind them. One such rumored animal stands out in particular from the above assortment, in that it widely is accepted as real. Adding credence to its possible existence is the uniformity of descriptions and testimonies regarding its morphology and its behavior.

Snake-That-Flies

Among rural Navajos, both traditional and modern, there is enduring recognition of an as-yet unclassified and unrecognized species of reptile, known as *Tł'iish Naat'Agii*, "Snake-That-Flies." Briefly, it is described as a snake (or at least as a snake-like reptile), possessing the actual ability to fly, as opposed to simply gliding, through aid of membranes or expanded skin extending behind the head and fanning out along the body, not unlike the exaggerated display of a cobra's hood.

Initially it may seem unlikely for such an animal to escape detection by science. But such reported reptiles are considered by the local people as a natural presence among the indigenous fauna. During my discussions with area residents, interestingly some of them were puzzled or became confused by an outsider's special attention to this particular animal. Knowledge of "Snake-That-Flies" by no means is limited only to the Navajo. I now understand that such creatures likewise are known to members of the Hopi Tribe, according to one source, as "Sun Snakes." While looking into this matter and also keeping my research colleagues informed of my pursuits, it was inescapable that comparisons would arise with descriptions and traditions that testify to flying serpents from elsewhere in North America.

At present, I've spoken with more than a dozen individuals who actually had sighted a specimen themselves, or at least who knew of a close family member, friend, or acquaintance who had experienced such a sighting. Aside from direct, second-, and third-hand testimonies, I also sought out and familiarized myself with the local lore attributed to the animals, which time may or may not prove accurate to the nature of the creatures. When speaking with witnesses I've taken great pains not to 'lead' or 'suggest' details or descriptions or otherwise to risk compromising the accuracy of the testimony. And while knowledge or awareness of the serpents almost is common in certain areas, it would not appear to be a matter of common-knowledge so as to be openly discussed and widely understood. It's been difficult in a number of cases to get people to divulge what they know, partly due to the traditional Navajo perception of snakes generally being negative, if not perceiving them as outright evil. And while some regard the serpent as a mere snake despite its extraordinary characteristics, others sometimes regard them as supernatural entities and therefore not a topic acceptable for open conversation.

The sightings in this report are by no means the extent of the accounts I've collected, but I've selected those that comparatively offer more in describing particular characteristics.

Most sightings involve the snakes traveling steadily through the air, though some accounts cite gains and changes in elevation, sometimes drastically so, confirming their time in the air is not limited merely to gliding. Most notable are four separate accounts that involve the animals achieving and then maintaining a circular course.

- Workers building a hogan were startled one evening when a "snake" flew inside the incomplete building and flew madly about in a circle, seemingly unable to find its way back out. The workers cautiously tried to kill the creature before it finally escaped.

- Two women related to me a memorable teenage experience they had sneaking out of the Tuba City high school dorms late one night. They made their way through a canyon and had arrived at a reservoir. A hissing noise suddenly was heard, seemingly overhead, which they compared to that made by a jet engine. Through the moonlight they saw a snake-like object diving down toward the reservoir over and again before lifting back up into the air. The object swooped down at the water until it apparently noticed the girls as they sought to hide behind a bush. Much to their horror, it then flew in circles above them, as if it was aware of their presence. Eventually it flew off and the two fled back to their dorm, feeling much chastened.

- A story from the 1930s describes how a Big Mountain area family held a bonfire after a day of tending their crops, only to have a "dragon" appear above them and fly in circles around the flames.

- In the 1980s, a teenage boy was reported as having shot a flying snake that swooped down at him whenever he neared its dwelling.

Propulsion

The mechanics of propulsion remain the greatest mystery of an already mysterious creature, though all reports coincide in citing how the entire body is in motion during the animal's flight. The puzzling aspect of the described motion is that it frequently is described as being exactly that of a snake slithering on the ground. One could assume that these exact motions only are being compared in memory to the more standard spectacle, but I've encountered firm insistence that the contortions in every way were like those of a typical snake, only differing by occurring in mid-air. Such claims defy logic, however, as anything that becomes airborne and remains so must force air forward, downward, and aside for aerodynamic lift. The horizontal undulations of reptiles would be worthless in this manner. It is not impossible for a snake to undulate vertically to some degree but it is difficult to imagine one doing so long enough or vigorously enough to lift its own body weight.

As to the role of the membrane, I mostly have been under the impression it serves as a passive sail, directing and somehow forcing air downward to generate lift. One report, however, would suggest they hold a far more active function. In a story told to me since beginning this report, a woman reported a brief encounter experienced by her grandfather and uncle. The men were in the midst of building a shade when a flying snake landed atop a nearby branch. Perceiving it as a good omen, the grandfather reached out and seized the creature behind its head. While in the man's grasp, the serpent began to *flutter* its semi-transparent ("like plastic") oval wings at such a rapid beat they faded out of view (a la the wings of a hummingbird or bumblebee). The man then released the animal and instantly it shot up into the air and flew out of sight. The story also offered some insight as to the size of the snake. She explained that her grandfather was a tall man (over 5'11", at least) and while he held the snake at his chest level, its body still reached the ground.

Contortion

Two mentions have been made about flying snakes carrying sticks or twigs. One woman told that her mother saw a flying serpent on the ground that carried twigs by arching over them and pinching them between the bend of its neck and head. This certainly would be impossible in position and function, based upon everything known about snake skeletons, and it just may be that she misunderstood what her mother had tried to describe. Oddly enough, I have heard another flying snake being seen with a small bundle of sticks that were constricted in its tail! As absurd as these stories sound, they would appear to acknowledge and support the belief that the serpents build nests in cliffs, which will be examined later in this report.

SOUND PRODUCTION

Flight Noise

Often cited is a type of 'hissing' sound, separate from a snake's threatening hiss, which likely is made while the reptiles are in flight, in many cases even before being seen.

- In the 1930s or 1940s, a Big Mountain resident was out on a morning walk when he heard a growing hiss from off in the distance. His son, who related the story, likened his father's description of the noise as very similar to a jet engine, like the drawn-out blowing of air. The man looked around until he spotted a snake's outline in mid-air, sparkling in the sunlight as it seemingly slithered by overhead.

- The two women who recounted their nocturnal encounter near Tuba City described hearing first a hissing noise which they likened to that of a jet engine.

- In 1996 or '97, a Hardrock area teenager out walking near his homesite heard a snake's hiss and immediately froze in place. He stood unmoving and carefully looked about. The sound grew louder but he saw nothing on the ground. When he looked up, he saw a snake with thin extended transparent flaps "twisting" its way through midair before it finally landed in a tree.

Voluntary Noise

Further distancing the creature from standard snakes is the claim that "Snake-That-Flies" emits sounds other than the reported hissing, such as a kind of growl. The only specific example I can cite involved one woman's account of what she saw and heard when she was a young girl. While that specimen flew through the air, she recounted that it made a "nice" hooting sound, like "looht, looht, looht," each such hoot seeming to be in time to the reptile's side-to-side contortions.

BEHAVIOR

Habitat

A) Nests

Aside from a capability for flight, the dwelling aspect of the reptiles almost is as remarkable. The creatures are reputed to construct cylindrical nests composed of small twigs and erected like towers against cliff walls. At present I have been guided only to one such structure, a collapsed nest in White Valley between Pinon and Hardrock. I have been informed that there is an intact nest near Big Mountain and another is said to be along the edge of a mesa near Hotevilla. I personally examined the White Valley nest and am enclosing photos taken at the scene. I've spoken to Big Mountain residents who claim to have seen the "dragon" nest in their area but so far have failed to inspire anyone to lead me there, as they traditionally and purposely avoid the site. The mesa nest near Hotevilla supposedly is known to (but not exclusively by) an elder residing in Tuba City. Apparently he shepherded a flock of sheep for a Hopi man in Hotevilla and often has watched the nest and its airborne inhabitants from a distance using binoculars.

MORPHOLOGY

General Appearance

Despite the flying aspect, the general consensus among witnesses and others well familiar with the lore is that the reptiles are a form of snake, as is opposed to being a 'snake-like animal.' In some accounts the witnesses cite the sole difference between a snake and the creature they had seen was simply the fact that it was steadily traveling off the ground. Only in cases where the animal was seen at close range was the presence of 'wings' noted. Sizes of the creatures tended to vary anywhere between 8 inches long to 12 feet in length, while the most common response was that they were 'snake-sized.'

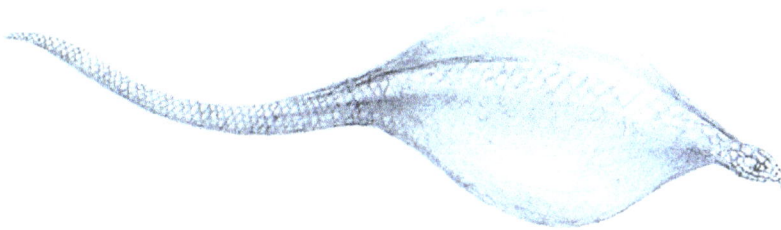

Witness' composite sketch (topside view), though the wings almost always are described as transparent, whereas in this image they are colored.

Coloration

Accounts vary where the animal's color and/or colors are concerned. This feature seemed quite secondary in reports by observers and, as most sightings take place from a distance, color seldom is discernible. However, one tribal elder told me the poisonous variant of 'Snake-That-Flies' could be identified by its "red belly."

Appendages

The "wings" undoubtedly are the most peculiar feature, even if they technically may fall short of the definition. Those only having heard of the snakes tend to assume they sport bat-like wings, whereas witnesses mostly have described something far differing. Instead of possessing limbs, the animals sport a retractable membrane that emerges from behind the head and then trails back to either side along a significant portion of the body length. Some have likened it to the expansive display of a cobra's hood, though both much larger and longer. This membrane is faint in color, almost to the point of transparency or, as one woman put it, "like they were made of plastic." Thus, it may tend to go unnoticed, creating the surreal image of a snake magically 'slithering' through mid-air. In three separate accounts that I have collected, one first-hand and the other two by second-hand, the membranes, or "wings," were noted as being "very beautiful" with a collage of "rainbow" colors. From these descriptions, such a spectacle sounds as if it only may be observable when light hits the membrane at certain angles.

Witness' composite sketch (bottom side).

BIOMECHANICS

That such an animal could persist so long beneath the radar of modern zoology perhaps is creditable to the simple fact that the concept of a *flying* snake likely is an oxymoron. In many cultures, including the Navajo, snakes are deemed dirty or vile due to their earthbound and slithering nature. Flight, in many ways, would be the very opposite of their lowly disposition. Symbolism aside, comprehending how a limbless animal actually could achieve true flight demands mental acrobatics. And that perhaps is the creature's greatest defense against discovery. Their very existence would defy not only basic concepts of herpetology but also our present understanding of bio-physics.

Flying

Officially, the only "flying snake" recognized today is a canopy dweller found in Southeast Asia. Its given title is unfitting, as the snake's aerial feats are limited to a partially-controlled downward glide. Its flattened ribs serve to make the animal ribbon-shaped, thus to slow the descent. Not since the pterosaurs of the Mesozoic Era are reptiles thought to have achieved legitimate flight. The possibility of a true *flying* snake here in northern Arizona has significant implications beyond being a "new" species. Its means of accomplishing flight entirely would be exceptional to all other (known) forms of flying animals. "Snake-That-Flies," if confirmed and documented, would be the only animal known to achieve flight without conventional wings.

Originally dubbed "Nest-That-Stands" and built entirely of small twigs, the structure, until relatively recently, stood up along this cliff wall in White Valley. Local tradition maintains the nest had been there possibly for 80 years or more. Sheepherders deliberately avoided the site on account of the flying serpent that inhabited the nest. Sometime in the 1980s, rocks above the structure gave way, collapsing the nest into its present position. (The 11-inch shoe seen is for size comparison.)

Exactly how snakes could construct such structures baffles one even more than trying to comprehend their flight without wings. Still, that they "make" nests is repeated over and again, even acknowledged in a Mexican reference that shall be included later in this. It may be that the serpents actually don't build these nests themselves but, more probably, come to inhabit derelict nests of predatory birds, mainly hawks or eagles. Snakes are opportunistic animals and will inhabit the underground burrows of owls or prairie dogs. However, as stated earlier, I've heard two separate references where the snakes allegedly were seen "carrying sticks".

As to the actual function of the nests and the number of occupants they hold, nothing I've gathered so far offers any clue. Based on local lore surrounding the White Valley nest, I have the impression that, until recently, it was a solitary snake that claimed residence. However, from a second-hand tale of the Hotevilla nest, it seemed as though a number of flying reptiles were observed emerging from the aerie.

B.) Rocks

More common than nest references are accounts or stories describing *Tł'iish Naat'Agii* residing within rock crevices. The serpents so far only have been cited in areas of rocky terrain. Inquiries about relatively flat areas, such as Teesto, always were fruitless. Stone enclosures may serve as transient homes as certain area stories speak of the serpents suddenly appearing but remaining in that area only for a limited time.

In the Big Mountain area is a conspicuous rock formation known as "Sitting Rock" to the locals. Memory and tradition maintain that it once was the dwelling of a lone flying serpent. High in the rock was a wind-eroded hole or 'window,' and the animal had settled within. Below the rock was a well-frequented footpath; both travelers and shepherds long had learned to keep their distance as the snake was reputed for its hostility.

CULTURAL SIGNIFICANCE AND PERCEPTION

The Navajo generally hold snakes in negative temperament and this has been something of an obstacle in my inquiries on the matter. If snakes as a topic initially are not considered 'too' taboo, then most times it seems that unprecedented focus on the subject begets hesitation. Whereas traditional Navajos regard snakes as lowly, filthy creatures, I've heard two references of offerings being made toward flying serpents or at least being left at spots where they were seen or were thought to dwell. Likewise, in the story explained beneath 'Propulsion,' the sudden presence of a flying snake was perceived as a good omen. Exactly where this perception originates or even how widespread it is among traditional Navajos remains unclear.

Of course, some encounters with *Tł'iish Naat'Agii* are received differently. A mesa resident of Red Lake told of a neighbor who, many years ago, was caring for her sick daughter. As was the custom when a child was sick, the woman brought the girl to a particular side of their hogan to bask in the sunlight. While they were there, a winged snake flew down, landed on the ground nearby, and immediately coiled its body. The woman apparently attributed the event as a sign pertaining to her usage of peyote in the Native American Church and, as the story goes, she became a pious Christian in response.

While Navajo elders' response to flying snake encounters may deem the experience positive in some manner, it would appear that younger generations hold no such reverence, as recent accounts end with the creatures being killed after having been sighted. One individual I contacted has concluded that the reason "Snake-That-Flies" is rare today simply is due to the persistent habit of killing the creatures upon discovery. If this is true, urgent incentive exists to identify it and recognize it as an endangered specie before it is too late.

Ethnology

If such animals have been present in the region for an extended length of time, as opposed to recent introduction or migration, then one should expect a form of acknowledgement within tribal belief structure or lore beyond mere anecdote. While I am completely unfamiliar with Hopi acknowledgement at this point, I've been told that the Navajo rituals for the Wind, Rainbow and Lightning Ways all make reference to flying snakes or at least make representation in sand paintings. I've been unable to confirm this, though I am aware of their role in the Bead Chant story. As told, the snakes descend from a sort of celestial dwelling to rescue the main hero who was left stranded in the clouds when his befriended eagle companions fail to carry him any higher into the sky. The snakes succeed where eagles fall short, triumphantly delivering the hero into the sky world.

There is a curious though painfully short reference to flying snakes in Gladys A. Reichard's "*Navaho Religion: A Study of Symbolism*" (Princeton Bolligen Publishing, 1950), which states on page 467:

> *One of Matthews' informants states that when the animals were people, the birds and snakes built cliff dwellings, and he asks the rhetorical question, "If they had not wings, how could they have reached their houses?" He explained why the snakes were able to help the Eagles and Hawks lift the boy to the sky in the Bead Chant story.*

OUTSIDE REFERENCES

Flying snake references are by no means exclusive to Navajo and Hopi lands or to the southwestern United States for that matter. Through the assistance of colleagues, I have been apprised of a small number of written accounts pertaining to the presence of, or the presence of a belief in, snakes capable of flight in other areas of the continent. Adding substantial credence to the matter is how particular attributes discussed in this report find echoes in these outside references.

Leon, Mexico

The first to be reviewed is titled *The Serpent* by Juana Pequeno and told by Nicole Ruiz, which was found by a research colleague at:
www.sanbenito.k12.tx.us/Schools/BertaCabaza/READING%20DEPT/Carmona/Snake/The_Serpent.html

> *My grandmother grew up in the mountains of Mexico, near a town called Leon. There was talk about a creature named "a serpe" or serpent. It was supposed to be a rattlesnake that had gotten very old. It had wings like a bat, feet like an eagle with long talons, and thick black hairs between some of the scales. Its head was just like a rattlesnakes with huge fangs.*
>
> *It built a nest like a birds, between large rocks. Supposedly, it was dangerous because it could bite like a rattlesnake, and it was easier to get you because it could fly. When it flew, it would make a strange humming sound. Whenever one was heard or seen, the villagers would chase it until they killed it. When my grandmother was eight years old, she was chased by a bull. To get away from it she climbed up on a big pile of rocks. When she reached the top, she saw a bird's nest and there in the middle was a serpent. As it started to uncoil, she jumped off the pile and ran to go tell her father what she had seen. Everybody from the ranch came back and killed it.*

Columbia, S.C., May 29. -- Closely following the appearance of the hand of flame in the heavens above Ohio comes a story from Darlington County, in this State, of a flying serpent. Last Sunday evening, just before sunset, Miss Ida Davis and her two younger sisters were strolling through the woods, when they were suddenly startled by the appearance of a huge serpent moving through the air above them. The serpent was distant only two or three rods when they first beheld it, and was sailing through the air with a speed equal to that of a hawk or buzzard, but without any visible means of propulsion. Its movements in its flight resembled those of a snake, and it looked a formidable object as it wound its way along, being apparently about 15 feet in length. The girls stood amazed and followed it with their eyes until it was lost to view in the distance. The flying serpent was also seen by a number of people in other parts of the county early in the afternoon of the same day, and by those it is represented as emitting a hissing noise which could be distinctly heard. The negroes in that section are greatly excited over the matter. Religious revival meetings have been inaugurated in all their churches, and many of them declare that the day of judgment is near at hand.

At this stage, it perhaps has become unnecessary to highlight the patterns of continuity that bind these unrelated references.

CONCLUSION

Pursuing evidence of "Snake-That-Flies" may have become a race against time. The majority to whom I've spoken on the matter believe the animals are on the decline, if not already extinct in some areas. What contributes to their demise, beyond human agency, is unclear. What is clear is that, without 'discovery,' without proper recognition, and without subsequent monitoring, the species very well may be lost even before it is 'found.' (All that having been said, now it just has been made known to me that there apparently was a sighting just last year near Shungopavi.)

Seeking out the creatures need not entail lengthy search and survey efforts into areas of previous sightings. If an open atmosphere were created that conveyed official interest in the animals, say, via the media, it might inspire those familiar with active dwelling sites to step forward and submit their knowledge of potential locations. Winter actually may prove the better time to seek this information, as traditionalists tend to discuss reptile matters more openly under reassurance that the creatures sleep, cannot hear their names, and therefore are unlikely to appear or to cause harm.

In addition to the significant implications "Snake-That-Flies" would hold for science, its 'discovery' would hold much benefit for the tribes, perhaps even as a tourist magnet. In turn, it might draw in a considerable amount of welcomed revenue, especially in the light of news that the Mojave station will be closing in two years.

Finally, the greatest implication such a discovery may merit is that, acknowledgement of such an animal would vindicate the rationality of traditional tribal knowledge and culture. Historically the tendency has been to dismiss native conceptual understandings of matters pertaining to nature, only to have many of them validated years later. The symbiotic relations of humans and animals, the medicinal values of particular flora, and even the intertwining complexities of ecosystems in the New World are just a few examples. As well, 'discovering' and proving such a creature, which long has been known to tribes of the Southwest, would serve to remind that Indo-European science has observed these continents only for five hundred years. And that there remains a lot of 'catching-up' to do with those native peoples whose devoted study of this land outstrips the scientific observations by twelve thousand years and more.

Nick Sucik
P.O. Box 497
Kykotsmovi, AZ 86039

(928) 380-1302

nicksucik@hotmail.com

"Phantom" Kangaroos in Germany
Ulrich Magin

This paper lists cases of kangaroos sighted, and hunted, and often caught, in Germany. In a majority of the reported incidents, a wallaby is either caught or missing, so we are dealing with real flesh and blood animals and not with phantoms.

My collection of cases is not the result of a systematic search, just an assembly of notes that have accumulated in the course of the last few years. There are years without incidents which simply means that I have missed them. Also, the German newspapers use the word kanguroo also to identify wallabies. Often, reports do not indicate any species. The collection is representative but not exhaustive. The sheer volume of case histories indicates how widespread the keeping – and escaping – of wallabies in Germany must be.

I have tried to find figures for the numbers of kangaroos privately kept as pets but failed – I only encountered internet pages with tips for keeping them, offers to sell them, and legislation how to treat them

For example:

♦ https://www.sueddeutsche.de/leben/exotische-haustiere-kaenguru-statt-schosshund-1.3915212
♦ https://praxistipps.focus.de/kaenguru-kaufen-das-sollten-sie-wissen_114151
♦ https://www.businessinsider.de/panorama/diese-exotischen-tiere-durfen-in-deutschland-als-haustiere-gehalten-werden/

That there is such a demand for tips how to feed, breed and raise kangaroos means that their number in Germany must be exceed by far higher any ideas that I have had about their numbers.

Before we go to that boring if informative list of German kangaroo encounters, I'd like to emphasise several points in more detail. The first is the question of naturalization attempts in Germany; the second is an overview of some reports or series of reports that merit special attention.

Naturalisation attempts in Germany
Although sometimes escaped kangaroos adapt to the environment and can remain in the forested areas for months (such as in the Taunus Mountains and in Mecklenburg) I am aware only of one attempt to naturalise kangaroos in Germany.

Around 1890, there were attempts to establish a breeding colony of Bennett wallabies for food production and hunting in the Kottenforst, a large forest area to the south and west of Bonn. Bennett wallabies are some 92 to 105 cm long – the males somewhat longer than the females, and weigh 14 to 19 kilo. The animals lived in Tasmania in a similar climate and did eat European hay, leaves, roots and bread, said local landowner Philipp von Boeselager. In 1887 he bought two males and three females and exposed them in the forest. In their first winter, the animals had to endure temperatures of minus 20° C (minus 4° Fahrenheit) which they survived without problem. "They are far smarter than any fox", remarked Philipp von Boeselager in a letter.

In the summer of 1890, traces of the wallabies were encountered in four areas of the forest, and it was assumed they came from nine individuals. In the autumn of the year, they had multiplied to 22.

And they found their end. Local hunters lay in wait at their feed-lots near Heimerzheim and exterminated almost the whole herd. Only two of the animals survived the slaughter. "It was many years later that we learned in which inn the poachers had eaten the animals", said Philipps' son Albert von Boeselager. One sighting of a wallaby was reported "by many eyewitnesses" shortly after from Brombach in the Taunus Mountains north of Frankfurt on Main, and it was assumed this was from the Kottenforst herd.

(Alfred Brehm: *Brehms Tierleben.* Vol. x, 4th ed., Leipzig and Vienna 1912, p. 211; Arbeitskreis Heimat Heimerzheim: *Heimatbote* #17, March 2012).

There had been several more attempts at feral herds of kangaroos in Germany around 1890 (of which I have no details), and a brief note I have refers to a colony which ended "after World War I" in the Taunus – so the Taunus kangaroo might have been one of them (*Natur und Museum*, 1943, p. 336).

The Manni Sightings in 1998
Some series of sightings have serious fortean overtones, like reports of kangaroos where none ever were, and the tendency of witnesses to see what they are looking for. Some of the reports I list later hint at such secondary sightings, but I have analysed only one such series of episodes, and here it is:

In Germany, Manni, a 3-year-old kangaroo, helped fill the silly season slow news gap in the summer of 1998 when it escaped from a private zoo in Bad Pyrmont (Lower Saxony) animal park on August 11. The animal was seen on August 12 20 km southwest of Hameln (Hamlin) in Aerzen, but it quickly vanished into the undergrowth. As soon as news of Manni's escape spread, people reported kangaroo-sightings from all over northern Germany. The animal park's director, Gerhard Grüne, said he'd received reports from as far as Schwerin (which is several hundred kilometres away). "Either these are hoaxes, or they can't tell a rabbit from a kangaroo", Grüne said (*taz*, 13 August 1998, p. 16 and 14 August 1998, p. 16; *Süddeutsche Zeitung*, 14 August 1998, p. 12). Additional sightings were noted from Minden in Westphalia, again far from Bad Pyrmont.

However, in the morning of August 13, it was observed near Groß-Berkel in Lower Saxony. Police were only informed four hours later and were unable to trace the escapee. This sighting was only three kilometres from where the kangaroo had been kept, so we can easily discount all additional reports. They cannot have referred to this animal (*taz*, 15 August 1998, p. 20).

Then the newspaper *Pyrmonter Nachrichten* reported that a motorist had seen Manni beside the A33 motorway, feeding on grass, at the Paderborn exit, some 80 km from Bad Pyrmont (*taz*, 21 August 1998, p. 20). At about the same time, the 2ft Bennet-kangaroo was also spotted at a Bad Pyrmont car wash, and a newspaper delivery man saw it at 6 in the morning at Aerzen, about half-way between Bad Pyrmont and the motorway sighting.

A woman came forward to say she had seen Manni on August 15, 1998, at the A 2 motorway near Porta Westfalica (40 km to the northwest) in the company of a herd of roe. "Well, all this is confusing", admitted Brigitte Schröter, the kangaroo's keeper (*taz*, 22 August 1998, p. 24). On August 18, 1998, an observation report came from Stade near Hamburg, 260 km to the north (*Bild*, 21 August 1998, p. 3). Then Manni was again seen close to Bad Pyrmont where it came from. A young man had seen the animal on a nearby mountain. "Manni was sitting on the road when a fox went towards him. Both animals than made it into the forest" (*Süddeutsche Zeitung*, 25 August 1998, p. 12).

The next sighting – or alleged sighting – was reported across Germany in its easternmost federal land, Saxony. A cab driver from Chemnitz saw it because she recognized "its long tail". Policemen arrived and were able to observe the creature through binoculars. At the same day, and only hours earlier (possibly on 29 August 1998) the kangaroo was also spotted near Bad Pyrmont (*taz*, 31 August 1998, p. 20). If police were right, there were at least two animals on the large.

At Erkelenz, which is close to the Dutch border near Cologne, police and firefighters were hunting a kangaroo for the whole day on August 30, 1998. In the morning, the animal was seen by policemen hiding in a cornfield, it was cornered in the evening and one huntsman was sure he had hit it three times with shots from a tranquilizer gun. A police spokesman for Heinsberg regretted at a press conference that the kangaroo had escaped, as "the darts had probably failed" (*taz*, 1 September 1998, p. 20).

In September, a psychologist explained that Manni sightings were due to extensive press coverage. At the same time, the Erkelenz kangaroo was still hiding in a cornfield at Kuckum, and attempts to chase it out with an ultra light plane had failed (*taz*, 2 September 1998, p. 20).

Meanwhile, it was reported that a kangaroo had also been spotted in Norway at the end of August 1998. It was presumed the animal had crossed over the border from Sweden, as no Norwegian zoo missed any (*taz*, 4 September 1998, p. 14).

The news magazine *Spiegel* summed up the affair on 7 September 1998, p. 268, implying cynically it all had been nothing more than mass hysteria (which was also my conclusion). The *Spiegel* added several additional sightings, namely one near Bad Wiessee in Bavaria and one on September 3, 1998, at Bad Pyrmont. However, on September 8, the Erkelenz kangaroo was caught with a shot from a tranquilizer gun and then sadly died of exhaustion in the arm of an animal keeper. So this at least had been real. On the same day, Manni was seen at 22 hours at the Maschsee near Hanover, again too far from Bad Pyrmont and also too late on the day for an average kangaroo (*Bild*, 9 September 1998, p. 3). On September 17, 1998, workers found the body of Manni close to the railway station of Bad Pyrmont. It had probably been lying there dead for a couple of days after a train must have hit and wounded it terminally (*Bild*, 18 September 1998, p. 5).

It is clear, however, that far more kangaroos had been reported to police and newspapers than actually were around. In the knowledge that a kangaroo was on the loose, people could interpret wildlife they saw as Australian exots.

The List, 1980–2020

1980s
I recall press reports from the early 1980s about a sighting of a whole herd of kangaroos in the Taunus mountains which claimed it was a breeding population that had lived there since the 1960s. But I didn't clip it then and I am unsure now as I haven't been able to trace this news report digitally.

1998
August: the Manni sightings.
♦ October: At Salzburg, Austria, a kangaroo escaped from a zoo on October 28 and hid in the forest (*taz* 30 Oct 1998, p. 20).

♦ October: On October 21, 1998 a newspaper delivery man at Finsterwalde (Brandenburg) found a kangaroo sitting in front of a supermarket door. It was obviously frightened and had actually just escaped from a zoo, to which it was easily returned (*taz*, 22 October 1998, p. 24).

1999

* January: On January 1st, 1999, six "dwarf kangaroos" escaped from a private zoo at Ostrhauderfehn in East Frisia. They were returned safe a day later (*taz*, 4 January 1999, p. 20). The article adds that another kangaroo had died at Mönchengladbach. Two further animals were returned safe to their zoos at Finsterwalde and Falkenstein.

* July: On 1 July 1999, a Tasman kangaroo escaped a private keeper at Rennsteig in Thuringia (*Süddeutsche Zeitung* 3 July 1999, p. 16; taz 3 July 99, p. 24). It had not been caught several days later (*taz* 5 July 1999, p. 20).

* October: According to the *Neue Badische Zeitung*, 30 October 1999, p. 5, a kangaroo escaped from a travelling circus on October 28, 1999, and was spotted close to the motorway A 81 by several witnesses. Police searched the area by car and helicopter, but by Friday so many observations had been reported at Böblingen-Sindelfingen that a police spokesman said he thought most of the witnesses had only spotted a rabbit.

2000

* November: A 9-month-old kangaroo escaped its owner at Heppenheim, Hesse, Germany, on 5 November 2000 and was not traced after. A zoo official said one kangaroo that had escaped in the Odenwald previously had lived for 2 years in forests and gardens until it was rammed by a policeman with his car (*Süddeutsche Zeitung* 9 November 2000, p. 16).

2001

• March: Five Bennett kangaroos escaped from a zoo at Burg Stargard near Rostock when somebody cut open the fence and literally flogged the animals out of their pen. The 2ft kangaroos were unsuccessfully sought by police (*Bild* 9 March 2001, p. 5). Two of the animals returned a few days later, a male and several female kangaroos, however, could not be caught and lived wild in the area (https://www.br.de/br-fernsehen/sendungen/welt-der-tiere/kaenguru-mecklenburg-vorpommern-100.html).

• March: Oscar, a Tasman kangaroo that had escaped in October 2000 in Thuringia, was caught in an Elementary School in March 2001 (*Süddeutsche Zeitung* 31 March 2001, p. 11).

2002

• February: On a Bavarian country road near Feuchtwangen, motorists spotted a kangaroo jumping across the road. It had escaped from its owner to whom it returned the following day (*Bild* 13 February 2002, p. 3).

2003
- May: Escaped kangaroo at Münster (*Bild* 12 May 2003, p. 6).
- September: A kangaroo escaped from the zoo at Kloster Ansburg in Hesse, was spotted by people in a garden at Nidderau-Windecken, and caught after two weeks near Friedberg, Hesse (*Bild* 5 September 2003, p. 3; 13 September 2003, p. 3; *Badische Neuste Nachrichten* 12 September 2003).
- September: After weeks in freedom an escaped wallaby was caught by policemen in the woods of eastern Belgium. Where it has escaped from was not known! (*Bild* 6 September 2003, p. 12).
- September: A wallaby escaped from a pet zoo at Gütersloh. It kicked and bit policemen severely who tried to capture it, but they finally succeeded. (*Bild* 10 September 2003, p. 3)
- October: A kangaroo escapes its owner at Wolfsburg, Germany; caught after 30 minutes (*Bild* 7 October 2003, p. 3).
- October: Skippy the 1.2m, 30kg, 2-year-old kangaroo caught 11 days after it escaped its owner at Magstatt-le-Bas, Alsace, France (*Badisches Tagblatt* 22 October 2003).
- November: Witnesses reported a kangaroo hopping across a Bavarian road to police who traced the owner at Veitsbronn. Also called Skippy, the kangaroo hopped into a train the next day and died (*Bild* 13 November 2003, p. 6; 14 November 2003, p. 3).

2004
- July: Early in July, a kangaroo escaped its private owner near Cham, Bavaria. It was seen two weeks later at Eggenfelden (*Bild* 14 July 2004, p. 6), and was caught there a few days later (*Bild* 17 July 2004, p. 6). At the very same time, witnesses reported a moose near Cham – *Süddeutsche Zeitung* 9 July 2004; *Badische Neuste Nachrichten* 9 July 2004.
- September: Early in September 2004, a dead kangaroo was found drifting in the Rhine at Grenzach-Wyhlen (*Bild* 7 September 2004, p. 5; *Süddeutsche Zeitung* 7 September 2004, p. 14).
- September: Roughly one week later, a 1.2m kangaroo was killed by a motorist when he ran over it near Singen (*sda* 14 September 2004; credit: Andreas Trottmann; *Neue Zürcher Zeitung* 15 September 2004, p. 43).

2005
- June: Three friends saw a kangaroo while driving near Bad Nauheim, Hesse, Germany. None was missing from zoo or circus (*Bild* 8 June 2005, p. 5).

2006
- August: A kangaroo escaped from its owner at Untermünkheim but returned the same day (*Bild*, Stuttgart 12 August 2006, p. 7).
- September: A kangaroo escaped at Erlangen, Bavaria, and later died from a

heart attack (*Bild*, 7 September 2005, p. 8).

2008

♦ June: Four wallabies escaped from their Farschweiler pen in May. The last of them was caught on 8 June in a garage at Osburg (*Trierer Volksfreund*, 9 June 2008).

2009

• July: And yet – we learn that the last of the four wallabies was still at large 14 months later and was occasionally spotted in the local forests of the Hochwald (*Trierer Volksfreund*, 16 July 2009).

• August: In neighbouring Palatinate, a Bennett kangaroo (*Macropus rufogriseus*) was also spotted on several locations: in Grünstadt and in Gerolsheim (where it took a rest in a vineyard and allowed a cameraman to take a picture) (Die Rheinpfalz – Unterhaardter Rundschau, 21 August 2009).

2010

• January: The last of the four escaped Farschweiler wallabies was possibly seen on 13 January on a federal road near Idar-Oberstein (some 40 km distance) by a 49-year-old motorist who informed police. Officers searched the area but found no trace of the animal (*Trierer Volksfreund*, 14 January 2010).

2012

♦ March: A kangaroo escaped from a private owner at Weilerswist near Bonn (*Heimatbote* 17, March 2012)

2014

♦ October: On 9 October, three kangaroos escaped a private owner at Cologne-Porz and were hunted by police (*Express*, 8 October 2014).

♦ November: A wallaby with several young escaped and was searched for in the region around Berlin (*B.Z.*, Berlin, 25 November 2014).

2015

♦ May: Police caught a kangaroo at Eschborn, in the Taunus, and returned it to its owner. A second kangaroo he was missing was caught only a day later (*Frankfurter Rundschau*, 1 May 2015; *Mannheimer Morgen*, 2 May 2015).

♦ July: A kangaroo observed near Brilon in the Sauerland, possibly an escapee (*wp.de*, 5 July 2015).

♦ July: After storms damaged a fence, a kangaroo escaped at Lünen, Westphalia (lokalkompass.de, 31 July 2015).

2016

♦ August: Another escaped kangaroo in the Sauerland, called Skippy (*Rheinpfalz*, 12 August 2016, credit: Joachim Hüther).

An Animal of a new Species found on the Coast of
NEW SOUTH WALES.

♦ December: Nine days after escaping from its enclosure, kangaroo Speedy was discovered and captured by a veterinarian. Speedy had escaped, with three fellow kangaroos, from a park at Münster, Germany. The first two animals were captured two days later, the third returned on its free will after a week. There had been reports that the fourth had been killed, as witnesses had seen its body, but luckily this was not true (*Kölner Stadt-Anzeiger*, 17 December 2016, p. 16)

2017
• June: A kangaroo spotted several times by motorists at Nohfelden near Wadern in the northern Saar (Trierer Volksfreund, 9 June 2017).
• November: A female motorist spotted a kangaroo on a road near Freisen in the Saar – close enough to St. Wendel to be the same individual as in the summer sightings (*Rhein-Zeitung*, 5 November 2017).

2018
• April: On a road near St. Wendel, and in the night, a motorist hit and killed a kangaroo. It had stood on the verge of the road between Walhausen and Wolfersweiler, accompanied by a second kangaroo which hopped away (Rhein -Zeitung, 16 April 2018).

2020
♦ April: At the end of the month, a kangaroo was seen by motorists at the side of the A7 motorway between Neumünster and Bordesholm in Schleswig-Holstein (*Holsteinischer Courier*, 27 April 2020; credit: André Kramer).
♦ July: Boomer escaped from its owner at Landgraaf in the Netherlands and was soon seen in many places in Germany along the border. It was observed in Herzogenrath on 14 July, two days later, fire-fighters managed to catch the animal (*Netzwerk für Kryptozoologie*, 20 July 2020).
♦ October: An escaped pet kangaroo in Thuringia (*mdr.de*, 14 October 2020).

No date
Several years ago, several red-necked or Bennett wallabies escaped from their owner in Mecklenburg and have lived in the wild since then (*http://www.heimische-tiere.de/Neozoen.htm*).

Conclusions
There is little argument over the existence or not of escaped kangaroos in Germany, so the conclusions are easily taken and brief:

1. In contrast to sightings of ABCs, most kangaroos get caught or return after a few days and are real animals, escaped from their owners.
2. In some cases, once kangaroos get newspaper coverage, sightings proliferate, even where the kangaroo has never been.

Analysis of an Alleged "Chupacabras" Mummy

Hans-Jörg Vogel / Berlin

Introduction: What is the "Chupacabras"?

A bizarre creature, often described as hairless with sharp claws and long fangs, feeding on the blood of other animals, is said to be gallivanting all over Latin America. The "Chupacabras" is a relatively new phenomenon of interest in Cryptozoology. Its name may be translated as "goat-sucker", derived from the Spanish words *"chupar"* (to suck) and *"cabras"* (goats). The stories revolving around the Chupacabras frequently mention its liking for the blood of chickens and goats, leaving behind two punctures, mostly in the region of the neck.

Descriptions of the Chupacabras differ greatly; in fact, they have hardly anything in common and sometimes even contradict each other. To this day, there are countless

(Artist's renditions of the Chupacabras © Julia Deppe)

versions describing this creature that has long entered the world of myths and legends. On the one hand, the Chupacabras is said to possess two sharp canine teeth to suck out the blood of its victims; other reports claim that it uses its long hollow tongue to do so, all of which is reminiscent of common myths about vampires. The creature, whose height is said to vary between 1 m and 1.5 m, is also described as carrying yellowish-green spikes on its back, which it can retract at will. The colour of its skin reportingly changes to adapt to its backgrounds and moods.

The Chupacabras became more widely known as late as the mid-1990s, when media in Puerto Rico, Chile, and Mexico first showed interest in the unusual activities linked to the legendary creature. Later on, there were further reports from the USA, Namibia, and a number of other countries. Whether a Chupacabras was responsible for the incidents in any of these countries remains entirely unknown.

Discussions about the existence of the Chupacabras are ongoing among German cryptozoologists, as well, dividing them into two camps: believers and sceptics. From a personal perspective, I am inclined to assume that the Chupacabras is nothing more than an urban legend that has gained ground rapidly. However, many questions remain without answer and various explanations unexamined to this day, paving the way for greater exploration of the topic in the future.

Within the scope of this article, I will refrain from addressing any such questions and instead focus on the analysis of an alleged Chupacabras mummy, which is currently in possession of the museum Tor zur Urzeit e. V. in Brügge (Schleswig-Holstein, northern Germany).

The Specimen: An X-Ray Closer to the Truth

I acquired the alleged Chupacabras specimen not long ago for our permanent cryptozoological exhibition at the museum Tor zur Urzeit e. V. in Brügge (Schleswig-Holstein). The specimen, provided by the seller along with the information that it was a so-called Chupacabras, whether real or not, seemed quite befitting for our exhibition.

The use of the specimen first required analysis and examination before it could be presented to the public as an exhibit. This, however, proved to be more difficult than anticipated. For whatever reasons, laboratories were reluctant to take on the project. German scientific circles, in general, tend to maintain a low profile when it comes to engaging with cryptozoological issues.

After countless rejections – sometimes there was no reply at all – from the institutions

(Chupacabras mummy – © Photograph Hans-Jörg Vogel)

(Chupacabras mummy – © Photograph Hans-Jörg Vogel)

and laboratories contacted here in Germany, it was time for a different strategy. I personally approached different contacts who could facilitate various individual tests of the specimen. An x-ray seemed to be a good start to create the foundation for further analysis. This led me to the nearby veterinary clinic, where nurses and doctors were undoubtedly astonished at my unusual request.

I consulted with the senior veterinary doctor, and, to my own great surprise, she did not object to an imaging test on the specimen. Costs for the x-ray were within reason and the image was available within a short period of time. The result was indeed astonishing.

The x-ray clearly showed that the specimen had been prepared using two different parts that did not naturally belong together. It was obvious that the left part of the mummy belonged to the body or skeleton of an animal, while the head section

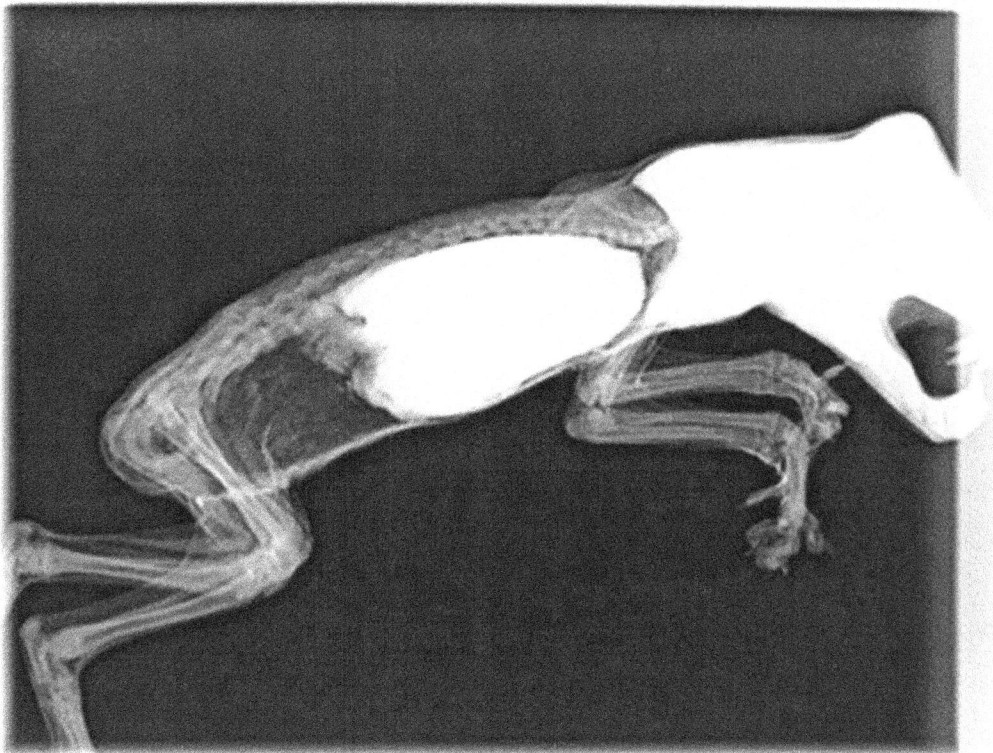

(X-ray – © SMARTVET / Berlin)

appeared to consist of a homogenous solid mass. The original head was, thus, missing. From the outside, the two parts were not distinguishable and appeared as one consistent structure.

In order to identify the animal to which the part of the skeleton had belonged, the veterinary doctor and I went through heaps of veterinary documentation and reference books. Given the lack of time, however, we were unsuccessful in our attempt. As per the doctor's preliminary assessment, the body part was feline.

Consequently, the next step was a closer examination of the artifact by a medical expert to determine the animal species. Markus Bühler, a medically trained and highly experienced member of the German Network for Cryptozoology, took on the task and was handed all photographs and x-rays.

Upper body and head

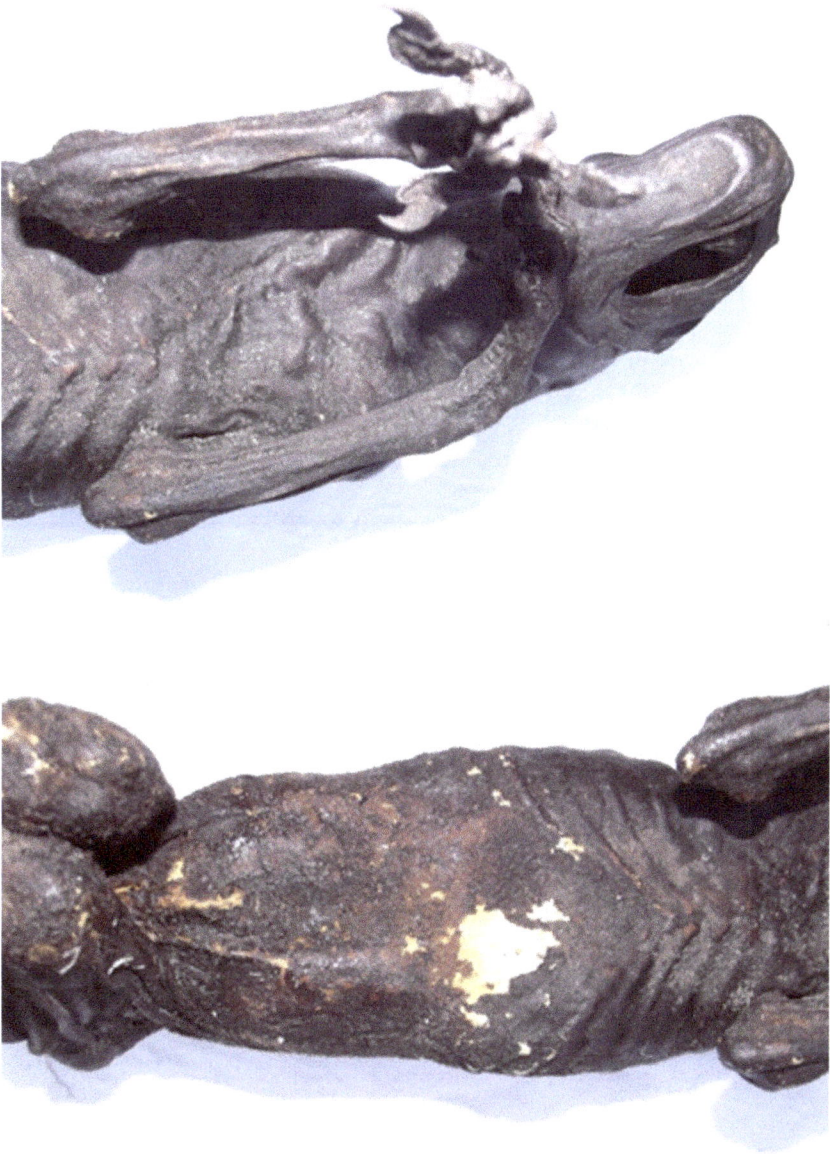

Underside

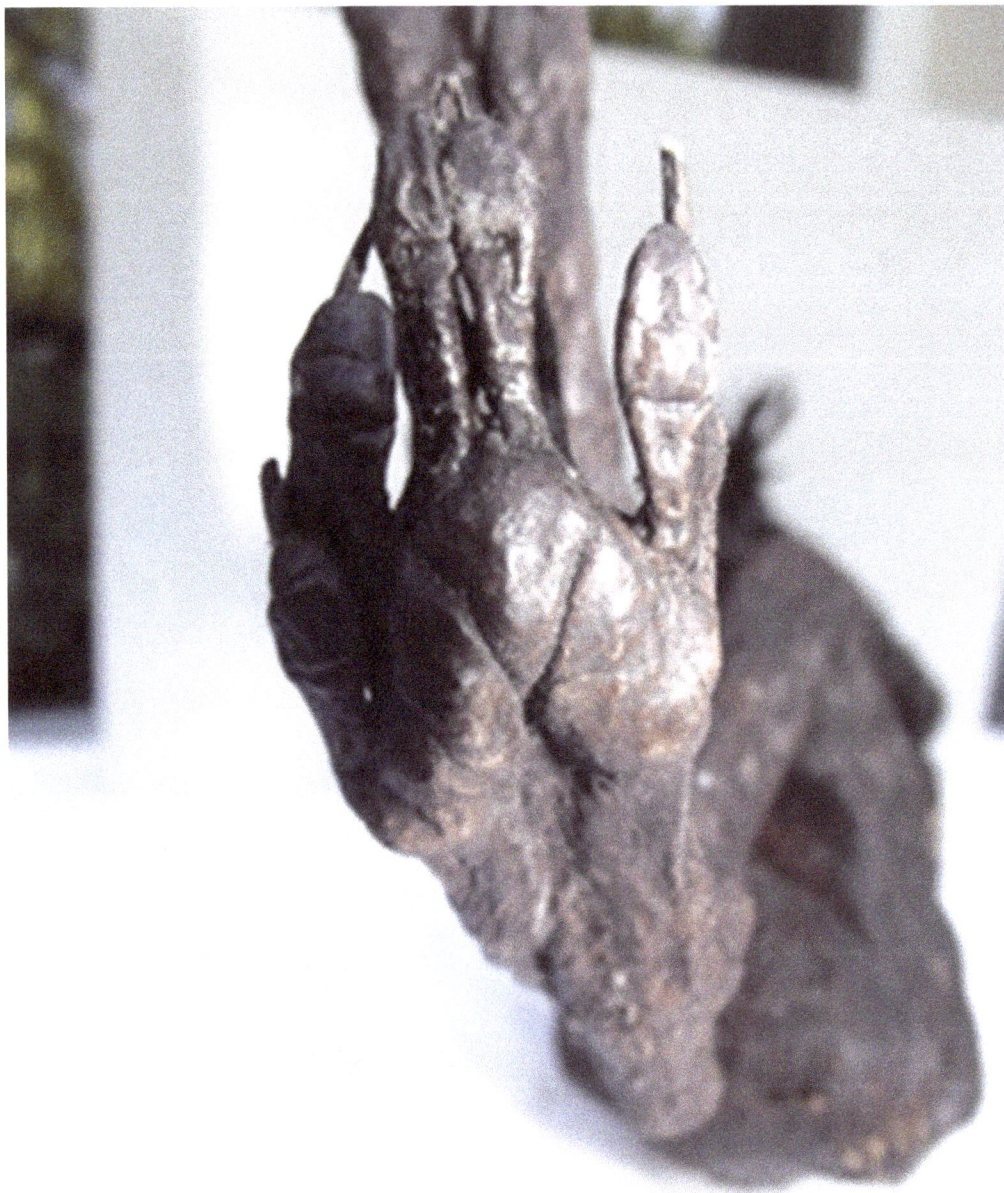

Hind limb

Forelimb

(All photographs © Hans-Jörg Vogel)

On the basis of the available digital photographs, Bühler was able to observe the following:

- *The skin on torso and limbs has dried up so that the underlying bones and some muscles and tendons are clearly visible; the skin appears very taut;*
- *Head and neck appear deformed;*
- *At maximum magnification, the photographs reveal that the skin's surface is partly covered with coarse grains of sand and a few thin white hairs, which have obviously been mixed into a paint or varnish, brushed over the skin to give a suitably rough surface effect;*
- *In the area of the abdomen, and partly on the limbs, this thin layer of paint or varnish has chipped off in places, revealing a light brown to yellow underneath, presumably the actual skin;*
- *The head, as well as the entire form, appears highly unnatural and in no way*

corresponds to an underlying skull;

- The eye-sockets are circular, and the nostrils are expanded oval in the most peculiar way;
- Of the ears, which have apparently been cut off at their bases, only a narrow circular rim remains
- At the front of the open mouth, two large canines as well as six small incisors are attached to the top and bottom respectively; no other teeth remain; the teeth, as well, show the mixture of varnish, grains of sand, and white hairs that has been applied to the rest of the body, which makes a more exact assessment more difficult;
- The limbs, on the other hand, remain in their original condition;
- The forelimbs end in hand-like paws with five very long toes which resemble fingers and feature small, pointed, and hook-shaped claws;
- The toes are arranged in one single line in the absence of dewclaws higher up on the legs that are typical of cats and dogs;
- The slender fingers do not have pronounced paw pads on the underside;
- On the other hand, the hind feet, likewise with five toes each, have well-developed soles, which reach to the heels, as well as clearly visible paw pads;
- It is therefore obvious that these are plantigrade feet, i. e. walking was done with all toes and metatarsals flat on the ground, indicating a plantigrade animal like a bear or a badger, as opposed to a digitigrade, such as a cat or a dog, which treads only with the toes of its hind feet.
- This finding is in line with the existence of five toes all located at the front of each foot; the pattern in which the paw pads were arranged is still clearly visible; notably, the first, fourth, and fifth toe (counted from the inside) each have one pad at the base, while the second and third toes are located together by the pad in the centre of the foot;
- Further examination of the torso reveals the absence of seams; no remains of external genitals are visible;
- The x-ray reveals a few other particularities of the mummy that are not visible to the naked eye, such as two large radiopaque shades that fill out the entire chest as well as the whole area of the head and neck; it can be assumed that this is a type of self-hardening modelling clay used to stabilise the body and to model the head and neck into the current peculiar shape;
- There are, however, no signs of a skull;
- Furthermore, it becomes clear that the mummy possesses indeed a tail, which has, however, been twisted and pushed entirely into the body – most likely through the anus – making it barely visible from the outside; the tip of the tail extends to the front end of the pelvis and measures about half of the length of the torso at full extension;
- A baculum, i. e. a bone found within the penis of certain male mammals, is not discernible, so that the body part in the specimen can be considered female.

**Racoon; source: Der Kryptozoologie Report
(Cryptozoological Report) no. 19**

Conclusion: A Man-Made "Monster"

All evidence indicates that the mummy is not a real "monster" but the body of an animal that has been manipulated in various ways to make it look as grotesque as possible. Unfortunately, the absence of the original skull, which might also be concealed by the radiopaque modelling clay, means that any identification must rely on other anatomical features.

The structure of the skeleton clearly shows that the specimen is a mammal, most likely a small carnivore.

Further discussions concluded with the identification of a possible "suspect".

This suspect is the Procyon lotor, better known as racoon. With a body length of 40 to 70 cm and a tail that measures approximately half of that, the mummy fits perfectly into the range of size of this species. The medium-length limbs, as well, are consistent with the proportions of the animal, taking into account that their fur makes body and limbs appear considerably more voluminous and compact than they really are underneath.

Most instructive, however, are the hands and feet of the mysterious mummy. Very typical of racoons are the long toes of their front paws which resemble fingers and have, in contrast to most other land-based carnivores, no developed paw pads. What is more, racoons are plantigrades and the arrangement of their hind feet's paw pads is fully consistent with what was found on the mummy.

The teeth attached to the front of the mouth could very well stem from a racoon,

(Photograph © Hans-Jörg Vogel)

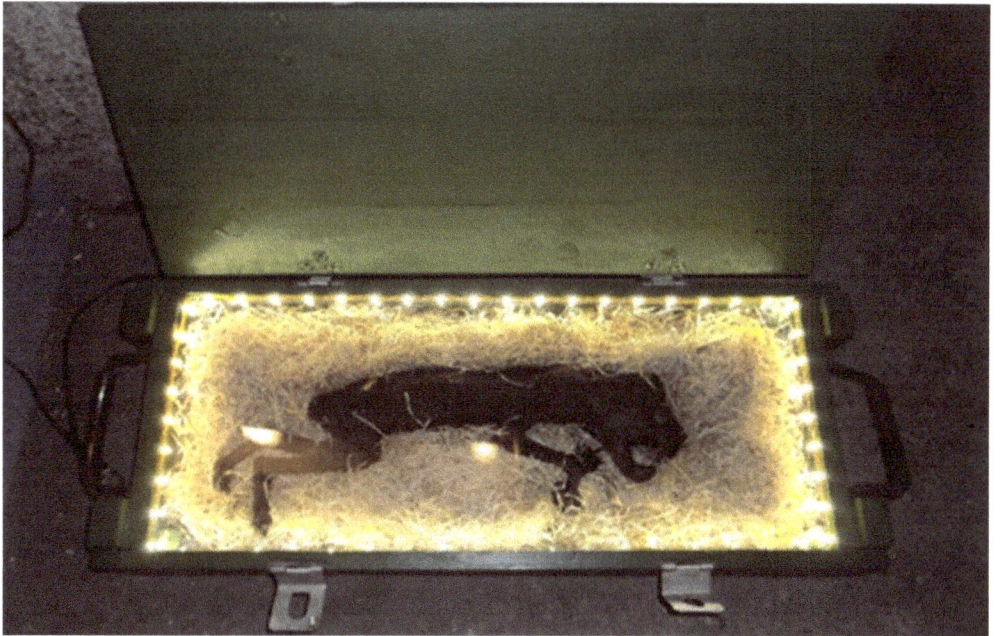

(Photographs from the exhibition - © Hans-Jörg Vogel)

DER KRYPTOZOOLOGIE

Report

AUF DEN FÄHRTEN VERBORGENER WESEN

Journal zur Kryptozoologie

Der Chupacabras

Mythos oder reales Wesen?

DER KRYPTOZOOLOGIE-
Report

Nr. 4
II / 2007

erscheint 3 x im Jahr, Preis: 10,00 € incl. Porto + Verpackung

DER KRYPTOZOOLOGIE REPORT

10. Jahrgang

Journal zur Kryptozoologie

Hauptthema

Ist das ein "Chupacabras"?

Eine Analyse von Markus Bühler

Literatur-Infos

und Rezensionen!

Der Kryptozoologie Report

Ist das ein

"Chupacabras"?

Röntgenaufnahme des "Chupacabras"-Präparates

Aufnahme: 21. Mai 1994

Wilde Kinder - Wolfskinder

Drachenformen in westdeutschen Städtewappen

Neun tote Skiwanderer am Toten Berg / Cholat Sjachl (Russland)

Der Djatlow-Pass-Vorfall

War es der Yeti?

VI. Kryptozoologie-Seminar

9. Mai 2015
im Museum "Tor zur Urzeit e.V."
und
Eröffnung der Teilausstellung

Kryptozoologie

Ausgabe Nr. 19   1/2015   Einzelpreis 8,50 € / 18,50 € in Farbe *although there is a theoretical possibility that they were taken from another carnivore of a similar size and dental status.*

This leaves to conclude that this specimen is with the greatest probability the mummified body of a racoon that has been disfigured in various ways. It is an excellent example of how even the body of an otherwise common animal can be manipulated in a way that makes is unrecognisable.

Following the results of the x-ray and the scientific examination, we deemed further analysis of hair and tissue samples unnecessary. This case of an alleged cryptid mummy was closed.

While this specific case will certainly not dispel the myth surrounding the Chupacabras, it highlights the absolute necessity for proper analysis and examination and, subsequently, the informed publication of the results. It is regrettable that much research in the field of cryptozoology is either not conducted at all or its results are kept from the public, a trend that contributes to the proliferation of speculation in the discipline.

The specimen has been integrated in our permanent cryptozoological exhibition in the museum Tor zur Urzeit e. V. in Brügge (Schleswig-Holstein, northern Germany).

Sources:

- Der Kryptozoologie Report 4 (II/2007): Der Chupacabras – Mythos oder reales Wesen? (The Chupacabras – Myth or Reality?)
- Der Kryptozoologie Report 19 (I/2015): Ist das ein Chupacabras? (Is This a Chupacabras?)
- Art. "Chupacabra", available online: <https://en.wikipedia.org/wiki/Chupacabra> (4 February 2022).

Further Reading:

- Markus Bühler, Bestiarium (Blog, EN/DE), available online: <https://bestiarium.kryptozoologie.net> (4 February 2022).
- Netzwerk für Kryptozoologie (Network for Cryptozoology), Official Website, available online: <https://netzwerk-kryptozoologie.de> (4 February 2022).
- Hans-Jörg Vogel, Der Kryptozoologie Report digital (Cryptozoological Report – Digital), available online: <https://www.der-kryptozoologie-report.de> (4 February 2022).
- Museum Tor zur Urzeit e. V. Brügge (Schleswig-Holstein), Official Website, available online: <https://www.torzururzeit.de/kryptozoologie> (4 February 2022).

The Modern Origins of Britain's Mystery Big Cats:

From WW2 to the Dangerous Wild Animals Act of 1976

By Tim Whittard and Matt Everett

The Battle of Britain

Imagine if you will; the year is 1944 and the rumbles and screams of enormous engines can be heard, as fighter planes litter the skies over RAF Ibsley in Hampshire. This sight is not an unusual occurrence; the 'Battle of Britain' is well underway and the collective efforts of the over-stretched allies are thrown into repelling the aerial invaders of the Luftwaffe. However, the fighter planes in question are not being flown by British pilots, but American ones; specifically, the 494th Fighter Squadron of the US Air Force, affectionately nick-named 'The Panthers' (The Wartime Memories Project, 2020).

So why is this relevant to the British big cat story, especially given that these "panthers" were not literal ones? The answer is simple; their mascots...

The insignia of the 494th Fighter Squadron featured a big black cat, a ferocious and menacing depiction of a panther, designed to instil a sense of confidence in the servicemen and women, whilst promoting fear in their enemies. This artful exercise in pinning a unique identity to the squadron did not remain the exclusive realm of the tailors and seamstresses tasked with embroidering the iconic emblems onto the uniforms of the servicemen and women; nor was it restricted to the painters and sign-writers who emblazoned the aircraft with colourful logos and designs. For in amongst the chaos of war and the mass movement of people heading east across the Atlantic Ocean to the battlefront in Europe, there were a (potentially large) number of soldiers and military personnel with unusual travelling companions.

Essentially, the point I am getting to is that a number of American soldiers are thought to have arrived in Britain, with pumas and panthers in tow. These majestic apex predators were essentially mascots, animal icons that served to inspire and motivate the soldiers who travelled alongside them; and yet, one important question remains...

Upon arrival on the shores of Britain during the height of rationing, and further beyond to the aftermath of the war, what became of these mascots?!

The truth is that records are at best sketchy, and at worst riddled with omissions and missing details; so the remnants of the story must be pieced together if any accuracy is to be garnered and helpful conclusions are to be drawn. Essentially much of the story of the big cats of Britain remains 'out there' to be told, but the details are scattered haphazardly in the old handwritten journals which were kept by homesick soldiers, many of which now serve as heirlooms in the private collections of families long bereft of their fallen war heroes (Whittard, 2020b).

However, that is not the end of the story and there are further layers of the metaphorical onion which can be peeled back in hopeful examination for more clues. One such notion which rears its head from time to time, focuses on the enforced rationing which was commonplace across and throughout Britain during WW2. Feeney-Hart (2013) reports on this very issue, describing how the British government advised its citizens to enact a cull of their animals in order to preserve food rations for the human population; this later became known as the 'British Pet Massacre of 1939', and in the space of one week it is estimated that 750,000 pets were killed. Newquay Zoo (2020) also point out that zoo animals were not always much luckier…

Reading between the lines it is clear that some owners of exotic big cats would have had very difficult decisions to make with the outbreak of war and the onset of rationing; the choice between feeding their family, or feeding their cats, the latter of which being obligate carnivores would have been unable to obtain the necessary nutrition to survive from any source other than a diet consisting purely of meat.

So there it is; the outbreak of WW2 may have itself directly contributed to the first wave of deliberately released big cats in the British Isles. Such animals may have been initially released from the pre-existing collections and menageries of exotic animals accrued during the previous decades and centuries of British colonialism (Whittard, 2020b). This may have seeded the initial nucleus of a new pseudo-population of big cats in Britain, which it seems were possibly further stocked by the arrival of military mascots from across the pond…

Working Class Gangsters, Upper-Class Eccentrics and the Exotic Pet Trade
For those of us who are intrigued by the frequent reports of big cat sightings in the wilderness and countryside of Britain, the Dangerous Wild Animals Act (1976) is likely to be familiar. It is an essential component of the tale, which is referred to by all enthusiasts and experts, but is actually understood in depth by very few. This piece of legislation is considered by many to be the very bedrock of the British big cat phenomenon - a notion we can be disputed to an extent, as mentioned earlier, but that said, its significance cannot be denied.

Everyone knows that the government of the United Kingdom introduced this legislation as a means of restoring order and control to the unregulated exotic pet trade of the British Empire which had continued as an extravagant fashion trend right into the mid-late 20th century. Many commentators will even be able to quote significant parts of the act word-for-word verbatim, and whilst some may agree that this demonstrates an in-depth understanding, it fails to capture the chronology of events which brought about its introduction; particularly the one key incident which was the 'straw that broke the camel's back', and resulted in the media spotlight being blindingly cast onto the exotic pet trade and a subsequent furore of debates in parliament which aimed to rectify any enduring legacy of irresponsible ownership of dangerous animals and the unregulated trade that ensured a steady stream of such creatures into our little island – in the interest of safety (Whittard, 2020a; Whittard 2020c).

The incident in question took place on Monday 26th October 1970, and judging by the few remaining written records which are easily available in the public domain, it must have been absolutely terrifying; on this day, eight year old 'James Tyler' (known as 'Jimmy'), was being dropped off by his father (a serving police officer) with his sister in the 'Oldfields Trading Estate' on Oldfields Road in Sutton, Surrey. Jimmy's father then left him and his sister to walk the short distance to meet their mother nearby, who was scheduled to finish work (Glenton, 1971; Brown, 2020).

Jimmy was unaware that a large male puma named 'Dax' was hiding under a nearby van (to which it was secured by a long chain, but with more than enough slack for it to be able to ambush, pounce on and maul an unsuspecting passerby); had he known of the puma's presence it is likely that Jimmy would have suppressed his instinct to run to his mother as he saw her exiting her workplace…

Alas, Jimmy did not know of the mountain lion that was waiting for him to approach, and nor was he able to defend himself from what happened next. Fortunately, he was rescued by a brave, fast-acting and quick-thinking man, who previously was stated to be 'unnamed'; however, I am now able to name the man as 'George McKnight' (Sussex History Forum, 2014). He was working nearby, witnessed the attack unfold, and without any hesitation stepped forward to the defence of Jimmy; using an iron bar to beat Dax into retreat, he was able to wrestle Jimmy free from the claws of the mountain lion, "before throwing James clear" out of the reach of the cat and thus ending the attack. Without the bravery and quick reactions of George McKnight, it seems incredibly likely that Jimmy would have needed a lot more emergency care than the short spell in hospital and the 38 sutures to his face, neck and throat that he did. Indeed, Jimmy Tyler owes his life to George McKnight.

Thankfully James was able to make a full recovery, and as far as can be told from researching this incident, he remains alive and well to this day; as does another interesting key figure connected to this incident…

The owner of the puma (or mountain lion) involved here is known by the name of 'Maurice Wheeler', (his full name is 'William Henry Maurice Wheeler'). In researching this incident and the history of Britain's exotic animal trade, the research team behind 'Britain's Big Cat Mystery' were surprised to be able to make contact with Maurice Wheeler, but less so, that any detailed discussions were halted as soon as the interest in the history of Britain's big cats was mentioned. Indeed from perusing the social media accounts attached to this evidently eccentric individual, it does not take long before the image of a wannabe hybrid between 'Peter Stringfellow' and 'Lenny McLean' emerges, with numerous photos depicting a life immersed in extravagance, wild nightclub culture, and imagery suggestive of gang-minded machismo; a view reinforced all the more by descriptions of his "dominant personality" and reports from investigations at the time of the attack on Jimmy Tyler, that "several people on the estate had witnessed various unreported incidents" associated with Maurice's mismanaged menagerie of man-eaters, but also that there was a well-established "reluctance on the part of witnesses to these events to commit their observations to writing" due to "fear of some form of retaliation" (Sussex History Forum, 2014). With this in mind, we suspect that action would have been a lot slower, if at all forthcoming, were it not for the fact that Jimmy's father was a respected and serving police officer at the time.

Maurice was known to keep several large cats at the time of this incident, and such was the concern about his ownership of dangerous animals in the immediacy of the following days, that factory workers on the 'Oldfields Trading Estate' went on strike, threatening not to return until Maurice's collection of big cats had been removed; eventually his exotic cats were re-homed by a zoo, but this was not the end of the issue…

This incident had caught the attention of the government, and only three days after the attack on James Tyler, a debate began in the 'House of Lords' to implement measures which would prevent such an incident from occurring again in the future (Hansard, 2020). It took nearly six years for this to be completed, but the end result was the passing of the Dangerous Wild Animals Act (1976) into British law, and the introduction of measures designed entirely to make domestic ownership of exotic or dangerous animals beyond the means of the average Joe. Logically then, it is fair to assume that this could have easily culminated in a wave of multiple releases of big cats across Britain by private owners who could not afford the licences required to legally keep their 'pets'.

Interestingly, the attack on Jimmy Tyler was not the first incident where a child was injured by one of his animals. Incredibly, just two months earlier, in August 1970, a 9 year old boy named 'Peter Cliff' was also attacked by another of Maurice Wheeler's 'pet' pumas; in this incident, Peter was passing the animal "when it jumped at him, gripped his shoulder and knocked him to the ground" near to a fishing lake in Dorking, Surrey (Glenton, 1971). What is even more unfathomable is that neither of these two incidents were the first, and that in fact, both were preceded by another traumatising event which occurred only 3 months before the attack on Peter Cliff, and 5 months before the attack on Jimmy Tyler. Unbelievably, on Friday 29th May 1970, the first recorded serious untoward incident occurred, when another 9 year old, this time a girl named 'Lorraine Wheeler', (Maurice's own niece) was bitten by one of his apes whilst she played in her own back garden!

It is often said by the learned and wise that 'those who forget history are destined to relive it' (Santayana, 1906), and clearly this sentiment rings true; as despite the Dangerous Wild Animals Act (1976) being decreed into law nearly half a century ago, problems persist with horrifying incidents of similar magnitude recurring. For example, in 2003, nearly three decades after the law was passed, an investigative journalism exposé revealed that the closed 'Basildon Zoo' had been selling lion cubs and leopard cubs illegally for as little as £700 through a 'black market' pet shop in Enfield, London which was ran by 'Steve Haswell' in conjunction with the zoo's then owner, 'Yolanda Surcouf' (BBC News, 2003; BBC Press Office, 2003).

As recently as August 2020, a video surfaced which showed a 16 year old girl being attacked by a captive puma, kept under license by 'Reece Oliver' at his farm in Strelley, Nottinghamshire (Gray, 2020); the girl (whose identity has been protected) was employed as a stable-hand, with no experience or training in caring for large carnivores, and yet on the day of the incident, she was instructed to work alone and unsupervised with the dangerous big cats, including the African lions, as well as the pumas - with terrifying results. Footage of this incident was captured on CCTV, and demonstrates how easily the puma was able to over-power its human keeper and escape the enclosure; which could have had even more disastrous consequences should the animal have made it beyond the perimeter fence (Pattinson, 2020).

It must be stated and emphasised that in all of the incidents outlined above, the animals involved were captive ones. This is of vital importance to remember, as whilst wild big cats do pose a threat to human life, it is markedly and drastically less significant than any risk posed by a similar captive animal; where the creature's entire existence is constrained to range-limiting enclosures and forced dependency upon human caregivers, who themselves, by the very nature of the responsibilities associated with owning a "Category 1" dangerous animal (such as a big cat) as defined by DEFRA (2012), are required to come into close contact with the animal on routine occasions, for provision of the numerous essential care and maintenance duties; with every moment of contact between human and any such captive big cat, being ripe with potential for catastrophé, given the slightest momentary lapse in concentration or common sense. Wild big cats would not naturally choose to socialise with humans and would exhibit an entirely different behavioural profile, which means that they would seldom, if ever, approach people and would prefer to avoid contact with humans in almost all circumstances.

Interestingly, it does seem that despite the vast majority of licence holders for the ownership of dangerous animals in the United Kingdom being extremely responsible individuals, there does seem to be an ongoing degree of confusion surrounding the legislation, which persists.

One could be forgiven for thinking that the Zoo Licensing Act (1981) does not apply to private animal owners, and yet upon closer inspection it is somewhat more ambiguous, stating that "the Secretary of State may from time to time specify standards of modern zoo practice" which must be adhered to; and upon reading the current revision of these standards, as outlined by DEFRA (2012), I was amazed to learn that the definition of a "zoo" can include private collections, "to which members of the public have access, with or without charge for admission, on seven days or more in any period of 12 consecutive months". By this definition, it is possible that any owner of dangerous animals who keeps them at their residency could find their 'private collection' classified as a 'zoo', should they have visitors to their home on more than 7 occasions in one year. This may seem pedantic, but it is in fact a serious distinction to make, because in inadvertently crossing into the classification of a 'zoo', the licensed animal owner suddenly becomes liable and responsible for a whole plethora of other contingency plans and safety precautions which are otherwise not mandated by law, which can include; comprehensive plans for a "response to an escape in all situations", such as plans for "recapturing the animal", "the provision of firearms and darting equipment to tranquillise or kill escaped animals", which in turn, requires "regular training with firearms and darting equipment", as well as an appropriately protected vehicle to aid the "recapture party", and the preparedness to euthanise any escaped animal.

It stands to reason then, that some private big cat owners will fall into a grey legal area, where neither the owners, nor the authorities are entirely clear of their responsibilities, requirements, and of the laws, standards and guidelines which may apply.

To summarise, here we have seen a small selection of the many incidents which highlight the failings of the Dangerous Wild Animals Act (1976) in firstly, restricting and controlling the sale of big cats (and other exotic animals) without a license, and secondly, in ensuring that avoidable attacks on humans from captive dangerous wild animals do not occur. We have also seen how easily the distinction between 'zoo' and 'private collection' can be blurred within the current legal frameworks. This is before even beginning to try to tackle the known (and rumoured) deliberate releases of big cats into the British countryside, and the domino-effect of issues that could arise from having a range of apex predators secretly living and breeding alongside humans on this little island.

Significantly, Shepherd et al (2014) state that whilst "large felid carnivore attacks cause exponentially fewer injuries than human conflicts, falls, or other environmental exposures, they have become a not infrequent and potentially preventable cause of significant human morbidity and mortality in the last several decades", before adding that communities of people "who live, work, and pursue travel and recreation at the urban-wildlands interface", as well as individuals "who may be exposed to the increasing numbers" of "felid carnivores" in both "captive" and "wild" settings, should be educated carefully with regard to the "risks inherent to these activities" .

So clearly, this is not going to be the end of the story; there will no doubt be further avoidable and untoward injuries resulting from human-animal conflict in both captive and non-captive settings, and there remains much work to be done if we are to even to begin to attempt to address the failings and pitfalls of the Dangerous Wild Animals Act (1976), as well as the unintended consequences of enacting this law.

Quite how to achieve this, whilst mitigating any risks to both the public and any cats that may be out there is another matter altogether...

REFERENCES

- BBC News (2003) Zoo accused over lion sale. [Online]. Available from: http://news.bbc.co.uk/1/hi/england/2780747.stm (Accessed on 2nd December 2020).

- BBC Press Office (2003) UK's Worst ... ? exposes illegal trade in wild animals. [Online]. Available from: http://www.bbc.co.uk/pressoffice/pressreleases/stories/2003/02_february/15/uk_worst.shtml (Accessed on 2nd December 2020).

- Brown, M. (2020) Has Anything Interesting EVER Happened In Sutton? Londonist. [Online]. Available from: https://londonist.com/london/best-of-london/has-anything-interesting-ever-happened-in-sutton (Accessed on 17th August 2020).

- Dangerous Wild Animals Act (1976) Her Majesty's Stationary Office. [Online]. Available from: https://www.legislation.gov.uk/ukpga/1976/38/pdfs/ukpga_19760038_en.pdf (Accessed on 2nd December 2020).

- DEFRA (2012) Secretary of State's Standards of Modern Zoo Practice. [Online]. Available from: https://assets.publishing.service.gov.uk/government/uploads/system/uploads/attachment_data/file/69596/standards-of-zoo-practice.pdf (Accessed on 6th December 2020).

- Feeney-Hart, A. (2013) The little-told story of the massive WWII pet cull. BBC. [Online]. Available from: https://www.bbc.co.uk/news/magazine-24478532 (Accessed on 10th October 2020).

- Glenton, G. (1971) A man's pet pumas savaged two boys. PRIVATE ZOO A MENACE JURY TOLD. The Daily Mirror. (Printed on 14th December 1971).

- Gray, M. (2020) Residents of English town want millionaire's big cats out after attack.

New York Post. [Online]. Available from: https://nypost.com/2020/08/29/residents-of-english-town-want-big-cats-out-after-attack/ (Accessed on 4th December 2020).

- Hansard (1970) House of Lords. Animals Bill. 29 October 1970. Volume 312. [Online] – Available from: https://hansard.parliament.uk/Lords/1970-10-29/debates/61d2a13e-8133-44a1-a828-7dc0ae4c4320/AnimalsBillHl?highlight=dangerous%20wild%20animals%20attacked#contribution-d8e3ef60-161c-43a7-96b2-95a84ddd6c2d (Accessed on 17th August 2020).

- Newquay Zoo (2020) World War Zoo – Growing Food for People and Animals in Wartime. [Online]. Available from: https://www.newquayzoo.org.uk/userfiles/userfiles/file/1_%20World%20War%20Two%20and%20Rationing.pdf#:~:text=During%20World%20War%20Two%20%28often%20shortened%20to%20WW2%29,tea%20and%20butter%20you%20could%20buy%20each%20week. (Accessed on 10th October 2020).

- Pattinson, R. (2020) Terrifying moment stable girl, 16, is mauled by mountain lion kept in private owner's UK back garden. The Sun. [Online]. Available from: https://www.thesun.co.uk/news/12527654/terrifying-moment-stable-girl-16-is-mauled-by-mountain-lion-kept-in-private-owners-back-garden/ (Accessed on 2nd December 2020).

- Santayana, G. (1906) The Life of Reason: The Phases of Human Progress.

- Shepherd, S., Mills, A. and Shoff, W. (2014) Human Attacks by Large Felid Carnivores in Captivity and in the Wild. WILDERNESS & ENVIRONMENTAL MEDICINE (25) p.220–230. [Online]. Available from: https://www.wemjournal.org/article/S1080-6032(14)00008-8/pdf (Accessed on 2nd December 2020).

- Sussex History Forum (2014) An unlikely happening in Sutton. [Online]. Available from: http://sussexhistoryforum.co.uk/index.php?topic=6554.0 (Accessed on 2nd December 2020).

- The Wartime Memories Project (2020) RAF Ibsley during the Second World War. [Online]. Available from: https://wartimememoriesproject.com/ww2/airfields/airfield.php (Accessed on 10th October 2020).

- Whittard, T. (2020a) On the Origins of the 'Dangerous Wild Animals Act' of 1976. Vocal Media. [Online]. Available from: https://vocal.media/petlife/on-the-origins-of-the-dangerous-wild-animals-act-of-1976 (Accessed on 10th October 2020).

- Whittard, T. (2020b) On the Origins of the Naturalised Big Cats of Britain. Vocal Media. [Online]. Available from: https://vocal.media/petlife/on-the-origins-of-the-naturalised-big-cats-of-britain-7q2a3d09fq (Accessed on 18th December 2020).

- Whittard (2020c) On the Origins of the 'Dangerous Wild Animals Act' of 1976 – PART II (Those Who Forget History are Destined to Relive it). Vocal Media. [Online].

Available from: https://vocal.media/petlife/on-the-origins-of-the-dangerous-wild-animals-act-of-1976-part-ii-those-who-forget-history-are-destined-to-relive-it (Accessed on 23rd December 2020).

- Zoo Licensing Act (1981) Her Majesty's Stationery Office. [Online]. Available from: https://www.legislation.gov.uk/ukpga/1981/37/pdfs/ukpga_19810037_en.pdf (Accessed on 6th December 2020).

AN EARLY VICTORIAN INVESTIGATION INTO ROMAN CRYPTOZOOLOGY

This article, which could be described as one of the earliest pieces of cryptozoological investigation, looks at an early example of Roman mosaic in the vicinity of Palestrina about 35km east of Rome. This mosaic attracted the attention of the antiquarian who wrote about it in *The Edinburgh New Philosophical Journal* October 1834-April 1835 which we reproduce here. This mosaic contained images of many mammals, birds, fishes, reptiles and plants, most of which were still alive in the early Nineteenth Century and are still around today. However several of the animals are interesting because they were not recognised either in the early Nineteenth Century or now. Alternatively they can be recognised but are out of place, such as tigers in Egypt or Ethiopia (these are the areas covered by the mosaic) or extraordinary in some way e.g a snake maybe up to sixty feet long. The essay has a few cryptozoological nuggets from closer to the Nineteenth Century such as mention of a race of white bears from Mount Libanus which, I think, is the same as Mount Lebanon. There are also animals in the mosaic indicating bears in Egypt or Ethiopia ,which is interesting as there are only known to have been bears in the Atlas mountains of Morocco and as far eastward as Libya. **RM**

III. On the different Animals and Vegetables which are represented upon the Mosaic on the Pavement of the Temple of Fortune of Palestrina.

Of all the antique monuments which afford a representation of a great number of animals and vegetables, there is none that is more curious than the Mosaic of Palestrina. Whatever origin is ascribed to it, and whatever may have been its object, certain it is that the artist to whom we owe it, has represented the various animals and vegetables with a precision and accuracy which cannot fail to command our utmost confidence. Before endeavouring to prove this, let us attempt to give some idea of the object of the artist in the construction of this beautiful and singular monument.

We find that this mosaic formed the pavement of the Temple of Fortune, in the ancient Præneste, a town of Latium, upon the ruins of which was built the town of Palestrina, distant about twenty-one miles from Rome. This mosaic, about 20 feet in length by about 15 in breadth, was placed in the sanctuary of the Temple of Fortune. As through the agency of the damp to which it was exposed, it was unceasingly wasting away, the Commandeur dal Pozzo caused it to be represented in eighteen drawings, which imitated the original colours; and Saurés, Bishop of Vaisar, gave a short description of it in his History of Præneste, printed at Rome in 1655 *.

Somewhat later the Cardinal Baberini, wishing to withdraw this mosaic from the accidents to which it was exposed, had it removed into the palace of the Princes of Palestrina †. It was

* Prænest. Antiquit. Hist., tom. i. cap. xviii.
† See the explanation of the Plate of 1721.

then engraved, in 1671, in the Latium of P. Kircher * ; and, in 1690, M. Ciampini published another engraving of it, which differed considerably from the former. It was on account of this discrepancy that the Cardinal Francis Barberini, grand-nephew of the former, caused the mosaic to be again copied in 1721 in new plates of a much larger size ; though some errors appear to have crept even into these. It was to rectify these that Montfaucon, and, later still, the Abbé Barthélemy, have both published delineations, in which this mosaic is represented in such dimensions, and with so much accuracy, that we can form determinate ideas respecting the different animals and ve-getables which are represented, points to which we shall solicit the particular attention of our readers †.

The different authors who have studied the purpose and ob-ject of this antique, seem to have inferred it from a passage which occurs in Pliny. According to this author, the mosaics deno-minated *Lithostrata*, were in use at Rome under Sylla ; and there is yet, he adds, to be seen at Præneste the one which he caused to be formed in the Temple of Fortune ‡. It has been concluded from this passage, that we must discover, either in the vicissi-tudes of fortune, or in the life of Sylla, marked references to the mosaic of Palestrina. Kircher has adopted with ardour the first of these opinions §.

But others have discovered in it Alexander arriving in Egypt at the time when the oracle of Jupiter Hammon had legalized his conquests ; believing that, under this emblem, Sylla de-sired to recall those oracles which warranted his own eleva-tion ¶. Others, among whom is Father Valpi, have seen in the figures no traits but those which characterize the Romans, and have chosen to believe that Sylla caused himself to be re-

---

* Latium Vetus Roman. 1671, p. 100. Also Monim., tom. 1. p. 82.

† Montfaucon, L'Antiquité expliquée, tom. ii. du Supplem. 1754.—Expli-cation de la Mosaique de Palestrine. Paris, 1760.—Mem. de l'Académie des Inscriptions, tom. xxx. p. 505.

‡ Pliny, l. 34. cap. 25. Ed. Harl.

§ Veter. Latium, tom. ii. p. 150 and 152.

¶ This opinion has been maintained by Cardinal Polignac, as may be seen in his *Dissert. in Calc. delineat. Edit. a* Card. Baberin.

presented *.   Others again, and Montfaucon is one, not being
able to recognise either the journey of Alexander or the vicissi-
tudes of fortune, have supposed that Sylla was satisfied with
there presenting the shores of Egypt and Ethiopia, and especially the course of the Nile †.

On the other hand, many of the learned have advanced very
different opinions.   Thus, in the eyes of the Abbé Dubos, the
mosaic of Palestrina is nothing more than a kind of geographic
chart of Egypt, or of the country watered by the Nile ‡.
Whilst, according to the Abbé Barthélemy, it was destined
solely to commemorate the voyage which the Emperor Adrian
made into Egypt.   Winckelman discovered in it the meeting of
Helen and Menelaus, a meeting which took place in Egypt
after the tragedy of Euripidus ; Chaupy, that it was intended
to represent the exportation of· grain from Egypt to Rome § ;
whilst, according to Nibby, its object was to represent the festi-
vities which they were in the habit of celebrating in Egypt
during the inundations of the Nile ‖.

Amidst all these contradictory opinions, there is evidently
agreement as to one fact, viz. that the scene represented upon
the mosaic of Palestrina was enacted in Egypt and in Ethiopia.
If there were any doubt on this point, that doubt would be ef-
fectually dissipated by a study of the animals and vegetables
that are found represented upon it.   This examination, therefore, will be found useful, not only as connected with the inquiry which now engages us, but also as it is an antique, which
antiquarians with reason have regarded as one of the most curious which have been preserved, and one which Poussin has in
part copied in many of his pictures.

Another particular which seems to be unanimously established
is, that the under part of the pavement of this mosaic, that
which lies to the north, has reference especially to Egypt.   This
is easily determined, not only because we see the Nile repre-

---

* *Veter Latium*, tom. ix. p. 51.

† *L'Antiquité expliquée.*   Supplem. tom. iv. p. 148.

‡ Reflexions Critique sur la Poésie, tom. i. p. 347.

§ Maison de Compagne de Horace, tom. ii. p. 301.

‖ Il, Tempio della Fortuna, Prænestiana Roma, 1825, p. 12.

The third kind, placed on the extremity of a rock to the right, and quite close to the preceding variety, very much resembles the Magot (*Simia sylvanus*, Lin.), which is still to be found in Barbary. Its head is narrower than that of the preceding race, and the hair which covers its body is also-more abundant.

The fourth variety, which is likewise placed at the end of a rock, and below the magot, at the same side of the mosaic, connects itself with the Papio (*Simia sphynx*, Lin.) This baboon is characterized by a prolonged snout, which is, as it were, snipt off at the point, where the nostrils are placed; and which gives it some resemblance to the muzzle of a dog. The name which designates this variety in the mosaic, has now altogether disappeared; but M. Saurés assures us, that he had seen it on the drawings of the Commandeur Dal Pozzo. These drawings represent, in the same fragment, the tiger, the land crocodile, and an animal named Σατυροκ. It is extremely probable that this name was not accurately decyphered, and that it should have been read Σατυρος, a name which remarkably corresponds to such an ape as that now under our observation. It is known that naturalists have reserved the name *Simia satyrus* to the orang-outang, probably on account of its formation.

The last variety of ape which is represented in this mosaic is the Tartarin (*Simia hamadrius*, Lin.), which inhabits Arabia. This ape is at the left of the antique, and above the lion. Near to him is written the word Κημπν, which comes near to the Κηβος, or Κηπος, or finally Κοιπος, by which appellations the ancients designated a kind of ape with a head not unlike that of the lion. It is perhaps on account of this circumstance that this ape has been considered by Montfaucou as a peculiar species of lion. At the same time this respected antiquarian is astonished, and, according to his view, with reason, that the able artist to whom we owe this mosaic, has placed such a dreadful carnivorous animal upon the branch of a tree. But his astonishment would without doubt have ceased, if he had recognised that this animal was nothing more than an ape with a great mane, and a tail as long as a lion's, to which, therefore, it had some kind of resemblance.

II. Feræ.

A. *Plantigrada.*

The first animal we refer to the order Feræ, is evidently a *bear*, with a short body, thick limbs, and a short tail. This bear, placed in the upper part of the mosaic, above and to the left of the animal named Ταβους, is there designated by the name of Κροκοττας, or Κροκουτας. According to Diodorus Siculus (lib. iii. p. 168), and Pliny (lib. viii. cap. 21), those animals named *crocottar*, which, according to them, combined the nature of the wolf and that of the dog, were found in Ethiopia. It is certain that the bear was very common at Rome ; for Caligula caused four hundred of them to be destroyed in the Circus, with nearly an equal number of panthers. It was for long a matter of astonishment how Ptolemy, in the celebrated entertainments which he gave in honour of his father Ptolemy Soter, caused a considerable number of white bears to be killed, because it was not known that this variety occurs any where else than in the frozen seas. But since Ruppel has found this species near Mount Libanus, we can understand how Ptolemy could easily have brought them from that region. It is at the same time true, that Dionis and Pliny himself, have given a different origin to the animal which we refer to the genus bear (*Ursus*, Lin.), without at the same time being able to determine the variety with any kind of accuracy *.

B. *Digitigrada.*

The first of the Digitigrada of which we shall speak, according to most writers, represents the otter (*Mustela lutra*, Lin.). The general form of this terrestrial quadruped, which besides holds a fish in its mouth, further demonstrates the correctness of the opinion. There are two of them, placed in the upper division, near two fresh-water tortoises. Near these otters we read the word Ενυδρις, or Ενυδρις, a name which was common to the otter, and to a kind of serpent.

Herodotus has also spoken in many parts of his works of these aquatic mammalia. He says (lib. ii. cap. 72) that these

* Dion. Cass., lib. lxxvi. p. 360 ; and Pliny, lib. viii. cap. 30.

animals, named *Enhydris*, were looked upon as sacred by the
Egyptians; and he adds (lib. iv. p. 9) that they were taken in
the marshes, along with the beaver and other aquatic animals.
After this authority and that of Montfaucon and the Abbé Barthelemy, there can be no doubt the artist who represented these
animals wished to depict the otter.

The second of the Digitigrada to which we shall direct our
attention is of much more difficult classification. This terrestrial quadruped obviously belongs to a Digitigrada, characterized
by long ears, and a tail not less conspicuously long, characters
which perfectly correspond to the dog kind, so that it is very
possible it might be referred to an animal of this class. But
we should not be surprised if the artist wished to represent
the black wolf (*Canis lycaon*, Linn.) a variety which still
inhabits Europe. Near this animal we read the words, Κροκοδιλος
χιρσαιος, which is to be translated " Land crocodile." Since
this writing does not appear to have been displaced, it connects very well with the animal it indicates. According to Herodotus, land crocodiles exist in Africa; and are met with in
the rivers. These animals were all more than three cubits long,
nearly five feet, and their ears were very large. Besides the
true water crocodile, which is figured upon the antique, and
which has no kind of resemblance to the terrestrial animal which
now occupies our attention, is constantly designated by Aristotle under the name of Κροκοδιλος ποταμιος.* From this we perceive that the ancients have delineated very different animals
under the common name of Κροκοδιλος, adding thereto epithets
proper to distinguish them from each other. It is thus that the
panther or the leopard has received the denomination of Κροκοδιλος παρδαλις, a name which serves to distinguish it from the red
wolf, and from the true crocodile (*Lacerta crocodilus*, Linn.).

The terrestrial quadruped which is placed at the uppermost
part of the mosaic, and which is pursued by Ethiopian hunts-
men, is the civet-cat (*Viverra civetta*). This animal, according
to the ancients, a native of Egypt, still inhabits the hottest parts
of Africa.

The mangouste of Egypt (Pharaoh's rat), so famous under

---

* Hist. Anim. lib. ii. cap. 10.   Also, De Part. Anim. lib. iv. cap. 2.

the name of ichneumon, is represented in the mosaic of Pales.
trina. It is found placed under two camelopards. It is figured upside down, probably on account of the displacement of
the fragment on which it was designed. This variety ought,
as it appears to us, to be referred to the *Viverra ichneumon* of
Linnæus.

We shall next mention the hyena, which, with a lion, may
be seen at the superior extremity, and towards the left of the
mosaic. Near these two animals is written the word Θυαντις, or
Φυαντις, and not Ωαντι, as it is printed in the engraving of 1721.
" One might believe," says the Abbé Barthelemy, that these
animals represent a kind of wolf-lynx ;" but this conjecture is
contradicted by the form of the name and by the figure of the
animals, which rather represent, the one to the left a hyena,
and the one to the right a lion. Besides, we must not forget,
as we have before remarked, that the animals beneath which we
now read the word Θυαντις, at a previous period formed the
same group with the animal called Ηονοκινταυρα, which is now
found at the opposite side. From this it happens, that the
word Θυαντις having been displaced, it can teach us nothing concerning the true names of the animals, which according to their
form have the greatest resemblance to the hyena and the lion, to
which we have referred them.

Underneath, and to the right of the two camelopards, is to
be seen a lionness and her cub, under which is written the word
λεαινα. There can be no doubt as to the animal to which this
appellation refers, and all commentators are agreed upon the
point.

On the right of the mosaic, and near to the sheep, are to be
seen two great carnivorous animals, near to which is written
the word Τιγρις, a name which would express the tiger. The
Abbé Barthelemy has found no difficulty about it, and has regarded these animals as truly tigers. We do not, however, see
how we can adopt this opinion, for these mammalia do not present upon their skin regular black bands, but spots of the same
shade, disposed in the most irregular manner. No more can
these animals be regarded as the panther or the leopard, which
have no bands, more than the tiger, but only spots.

There is another carnivorous animal, of the tribe *Felis*, which

represents an animal, underneath which we read the word
Δγίλαρς, or Δγιλαρτ, according to the Abbé Barthelemy; and
Αγιλαρου, according to Montfaucon.   This word, so read by
these two commentators, is not to be found in Sauré's description of the mosaic.   Barthelemy remarks, that as the word is
situated upon the edge of a fragment, it has probably suffered
from the removal to which it has been subjected.   This able
antiquarian, being ignorant of its meaning, refers it to the animal near which it is written, which is an ape.   But the form of
its feet, not less than its other characters, oppose this idea.   It
is also to be observed, as Montfaucon remarked, that this animal must have been very formidable, since the mosaic represents
that several Ethiopians are occupied in attacking it, armed with
spears and bucklers, whilst others are placed in ambush, to
shoot it with their arrows should it pass near them.   These observations, joined to the characteristics of this species, induce us
to consider it as having belonged to a formidable carnivorous
animal, as, for example, the guipard, or hunting tiger of India
(*Felis jubata*, Linn.), or perhaps to the black panther (*Felis
melas*, Peron.), or some other great feline species.

Finally, the panther, which is placed above the Ethiopians
who are wishing to strike the guipard with their arrows, and
above which is written the words Κροκοδιλος παρδαλις, is also so
well depicted, as to be readily recognised.   We have seen that,
with the help of various epithets conjoined to the word Κροκοδιλος,
the ancients have designated animals very different from the
true crocodile, which they invariably named Κροκοδιλος ποταμιος.
As, besides, this animal differs from that beneath which we read
the word Τιγρις, it is probable that this rather represents the leopard than the panther.

III. PACHYDERMA.

The animals of this family which are represented upon the
mosaic of Palestrina, belong to a very considerable number of
species, and these very important, such as the hippopotamus
and the rhinoceros.   There can be no doubt as to the first of
these animals, which is represented with really more accuracy
than upon the greater part of the other monuments of antiquity.
These animals are here designed entire, with the exception of a

single individual, of which we do not see much more than the head raised above the waters of the Nile. The characters of the hippopotamus are so well indicated upon the mosaic, that it cannot be confounded with any other animal.

As to the rhinoceros, it is not so well drawn; and as it has only one horn, it must be regarded as one of the Indian species. The word Ρινκερος, written underneath, still more indicates, if there had been room for doubt, the animal to which it must be referred.

Two other pachyderma, placed to the right of the rhinoceros, have the Greek word Εφαλες, or Εφαλες, written over them. This word is found at the margin of one of the pieces of the mosaic, and perhaps, in the act of moving, some of the letters had been lost. It is still, however, possible that this figure represented an animal which, according to Pliny and Solinus, was found in Ethiopia, where it was known under the name of " Eale " (lib. vii. cap. 21.—Solin, cap. 55.). This species, as to size, comes near to the hippopotamus; it was fawn-coloured; its tail resembled that of the elephant, and its jaw that of the wild boar. Its head was furnished with horns, which sometimes pointed downwards. The majority of these characters correspond with the pachyderma, and particularly with the wild boar, some of which have tusks so long as to resemble horns. The Latin word Eale, and the Greek word Εφαλες, differ only in the termination, and the addition of one letter, which was perhaps forgotten in the text of Pliny, or, more probably, added in the mosaic of Palestrina. However this may be, the animal to which the term Εφαλες refers is evidently an animal of the order Pachyderma, and of the genus wild-boar. All that remains, therefore, is to determine the species. A large tubercle, supported upon a bony protuberance, exists only in the Madagascar hog, the *Sus larvatus* of Cuvier; and this variety now exhibiting it, there is much probability that it is to this species that we ought to refer the animal under review.

The second species of the genus, near to which we read, according to Montfaucon, the word Χοιροποταμον, or, according to Barthelemy, Χοιροπιθηκος, is of much more difficult determination. According to the first of these antiquaries, the expression by which the artist who constructed the mosaic would have designated this boar, was the river-boar; whilst, according to the other, it simply signifies the pig-monkey, " owing, perhaps," he observes, " to the figure partaking of the nature of both these animals." But this animal has nothing of the monkey in it; it is wholly of the boar kind, and consequently has no relation to the species of which Aristotle speaks, and of which the head resembles that of the chameleon (lib. ii. c. 2). Barthelemy, also, has himself remarked that he has not been able to perceive this last resemblance as corresponding to the figure. This last species is characterized by a pointed and very long snout, by a heavy and thick body; it is low in its limbs, with a short tail, but little coiled upon itself, and especially, it has tusks which do not protrude from its mouth. These characters do not appear to agree with those of any of our present races of the boar, and therefore it possibly may belong to some lost race of this genus, or of some other analogous to it.

The mosaic supplies us with another Pachyderma, near to which is written the word Ξιθιτ, a name which is unknown to Barthelemy as much as the animal it is intended to represent. According to Montfaucon, on the other hand, the animals named " Xithit," were very common in Egypt; and, according to him, it was the same as the rhinoceros, denominated by the Ethiopians Ara or Harisi; so at least says Comas the Egyptian. If this animal ever existed, there can be no doubt it is destroyed; for now we are not acquainted with any Pachyderma with teeth that are pointed, long, and sharp. This species would even constitute a new and distinct genus, if all is true respecting the several particulars represented. It should be added, that the existence of this animal seems so much the less doubtful, inasmuch as Kircher, in describing the mosaic of Palestrina, observes that the animal named ΞιθιΤ is a boar which is famous in Egypt, because it is only found in the neighbourhood of the town of ΧιθιΗ. From this we may readily judge how easily this species, so circumscribed in its abode, might become extinct.*

* Latium, id est, Nova et parallela Latii, tum veteris, tum Novi, descriptio. Amsteldami, 1681, p. 100.

IV. SOLIDUNGULA.

Two species of Solidungula are figured upon the mosaic. The first represents the common horse (*Equus Caballus*, Linn.); whilst the second, under which is written the word Λυξ, seems to be a race which is lost or destroyed. The orthography of the name proves that the antique is to be dated in the first ages of the empire. Previous to this epoch it would have been written Λυγξ. The animal to which this name is erroneously attached appears to be a species of horse, between the dzhiggtai and the quagga. It has nothing in common with the lynx of the ancients, which was the wolf-lynx, as it has been well remarked by Perrault.* In fact, the slightest examination suffices to show that the animal named lynx in the mosaic has solid feet, or but a single hoof; with the body, head and tail peculiar to the horse. In conformity with these characters, this specimen then, is neither the dzhiggtai nor the quagga, and still less the ass or the zebra. According to this, then, it would constitute a species which is now lost; if this race has really existed with the form and the proportions which are bestowed on it in the antique. On this point we may again remark, that this is the more probable, since the figures of the mosaic are generally so well delineated as to lead us to conclude that they have been copied from nature.

V. RUMINANTIA.

Four species of the Ruminantia are found on the pavement of the Temple of Fortune; and they all belong to the Ruminantia with horns. The first is the camelopard, distinguished on the mosaic by the word Καμιλοπαραλι. According to Belon, Aldrovande, and Gesner, this animal received its name on account of its form and its skin; because that, with the head and horns of a stag, it had the neck of a camel, and a skin spotted like a leopard. Its tail was small, its feet very unequally forked, and its fore feet much longer than its hind ones. Its horns, which were at the upper part of its forehead, were not above

---

* Memoirs de l'Acad. des Sciences depuis 1666 jusqu'a 1699, tom. i. prem. part. p. 131.

six inches in length.* This description, in conformity with the mosaic, agrees too accurately with the camelopard to allow us to doubt that the artist had not designed to represent these animals, the most curious of Africa, and which had been brought on different occasions to Rome, in the triumphant processions, and for the games of the Circus.†

The second species, near to which we read the word Γαβους, is more difficult to determine. Let us first attend to what has been said by the two antiquarians who have given us an explanation of the pavement. Montfaucon observes that the last syllable of the word Γαβους signifies " bos," an ox. But as the name is Ethiopic, there is no propriety in dwelling upon the conjectures that this coincidence might suggest. Barthelemy, by adding to the first letter a limb which had disappeared, converts Γαβους into Ναβους. Under this denomination the Ethiopians designate an animal which, with the neck of the horse, the feet and legs like an ox, has a head like that of the camel. The reddish colour of the nabum, intermingled with its white spots, had led them to bestow upon it also the name camelopard. And thus, under this denomination, the ancient authors have confounded two distinct species, which the author of the mosaic has very well distinguished.

It results then, from these observations, that the animal named *Nabum* in Ethiopia, really existed in that country, and at an epoch which, if it was not cotemporaneous, was at least little distant from that of the artist of the pavement. But if this species be now lost, its destruction must have taken place within the times of the records of history. For where do we now find an antelope, or an ox, having a hunch on the anterior and superior part of its back, with short and straight horns like the camelopard, with a head like a camel, whilst the limbs re- ˑ

---

* Belon, Observat. cap. 49. p. 263.—Aldrovande, Hist. Quad. p. 927.—Gesner, Quad. tom. i. p. 147.—Dapper, Description de la Haute Ethiopie, p. 420.

† We may here remark, that the camelopard which is figured on the mosaic much more resembles that of the Cape than that of Sennar, which at the present time is living at Paris. This latter, as is well known, has a more finely streaked coat than the Cape one, and a form which more corresponds with the variety figured on the antique.

six inches in length.* This description, in conformity with the mosaic, agrees too accurately with the camelopard to allow us to doubt that the artist had not designed to represent these animals, the most curious of Africa, and which had been brought on different occasions to Rome, in the triumphant processions, and for the games of the Circus.†

The second species, near to which we read the word Ταβους, is more difficult to determine. Let us first attend to what has been said by the two antiquarians who have given us an explanation of the pavement. Montfaucon observes that the last syllable of the word Ταβους signifies " bos," an ox. But as the name is Ethiopic, there is no propriety in dwelling upon the conjectures that this coincidence might suggest. Barthelemy, by adding to the first letter a limb which had disappeared, converts Ταβους into Ναβους. Under this denomination the Ethiopians designate an animal which, with the neck of the horse, the feet and legs like an ox, has a head like that of the camel. The reddish colour of the nabum, intermingled with its white spots, had led them to bestow upon it also the name camelopard. And thus, under this denomination, the ancient authors have confounded two distinct species, which the author of the mosaic has very well distinguished.

It results then, from these observations, that the animal named *Nabum* in Ethiopia, really existed in that country, and at an epoch which, if it was not cotemporaneous, was at least little distant from that of the artist of the pavement. But if this species be now lost, its destruction must have taken place within the times of the records of history. For where do we now find an antelope, or an ox, having a hunch on the anterior and superior part of its back, with short and straight horns like the camelopard, with a head like a camel, whilst the limbs re-

---

* Belon, Observat. cap. 49. p. 263.—Aldrovande, Hist. Quad. p. 927.— Gesner, Quad. tom. i. p. 147.—Dapper, Description de la Haute Ethiopie, p. 420.

† We may here remark, that the camelopard which is figured on the mosaic much more resembles that of the Cape than that of Sennar, which at the present time is living at Paris. This latter, as is well known, has a more finely streaked coat than the Cape one, and a form which more corresponds with the variety figured on the antique.

semble those of the ox ? Such a species, which on the authority of the mosaic and the commentators is far from being monstrous, since it did live in Ethiopia, having been no where found in our days, we must thence conclude, that, like many other races, it has disappeared from the surface of the earth ; or, at least, that the interior of Africa, so far as it is yet known, has none of them.

The two other Ruminantia which are found in the Mosaic are well known. The first is the sheep ; and as to the second, which is led by a peasant along the banks of the Nile, it is evidently a representation of the common ox (*Bos taurus*, Lin.) Close to the sheep is seen the words Αρος, probably for Δορκος—the wild-goat. Notwithstanding, the animal near which the word is written is certainly a sheep. Barthelemy is with reason astonished that, in the engraving of 1721, they have substituted the word Απρος, a boar, which in no degree resembles the figure in the mosaic. This mistake is also fallen into by Montfaucon, though he could not trace in this animal any of the peculiar characters of the Aper or wild boar.

BIRDS.

Numerous birds are represented on the mosaic of Palestrina. They belong to three different families; 1*st*, The Gallinæ ; 2*d*, The Echassiers (Cuvier) (Grallæ, Lin.) ; and, 3*d*, The Palmipedes. The last are the most numerous, both as it regards species and individuals; and this arising from the mosaic representing the course of the river Nile.

I. GALLINÆ.

A. The domestic peacock (*Pavo cristatus*, Lin.). This bird has been represented on an immense number of antique monuments, often with its tail expanded, and in other positions.

B. The common pigeon (*Columba livia*, Lin).

II. GRALLÆ OR WADERS.

A. The white stork (*Ardea ciconia*, Lin).

This bird has been represented upon the top of one of the cradles, which is constructed of twisted rushes, which are to be found on the mosaic, probably with the intention of pointing

ter ϱ is probably left out, which would make it Σαυϱος, signify-
ing a lizard.

III. Ophidia.

The largest of the Ophidia which is figured upon the mosaic
has been considered by all commentators as the giant serpent,
so named, they state, on account of its enormous size.   In truth,
according to Diodorus Siculus (lib. ii. p. 149; also lib. iii. p. 169;
and also lib. i. p. 29), it existed to a very great size in Ethiopia,
and also in the islands which are formed by the Nile.   But the
question occurs, To what precise species are we to refer this gigantic serpent ? It is unquestionably of the python or boa kind,
genera in which we find the largest known species.   As respect-
able naturalists have affirmed that the serpents to which the
name boa has been given all came from America, if this be true,
the species represented on the mosaic must necessarily be one of
the great pythones of Africa, similar to that which Augustus
exhibited at Rome in the games of the Circus, and which, they
assure us, was sixty feet long, or like to that which was besieged
by Regulus' army.

Finally, the last of the Ophidia, which is found on the mosaic,
was known to the ancients under the name of *Ophilini*.   It appears that it may be referred to the variety Haje, or the *Vipera
haje* of Geoffroy de St Hilaire.   It is known that this species is
still frequently met with in Egypt, and that jugglers have the
art of taming it.

There are still some other animals represented upon this antique of Præneste ; but as they are invertebral animals, and
more particularly crabs, regarding the classification of which
there can be no great certainty, we think we need not say more
upon the subject.   However, that we may still exhibit the care
which the artist has taken accurately to copy the different objects he has introduced into the picture, we shall say a few
words on the plants that are found on it.

We may, in the first place, remark, that these vegetables
have been already recognised by M. de Jussieu, whose very
name carries authority with it.   At the side of the porch, where
the Emperor Adrian is standing, we observe a cocoa-palm tree
loaded with its fruit.   Behind the porch there stands a juniper

tree, between cedars; whilst near the portico, where the priests are, there is another individual of the same class. Regarding the tree which is placed near the great round tower, nearly in the centre of the mosaic, it evidently belongs to the Cassia family; and those which are seen in the same row, running to the right of the mosaic, are date-palm-trees. This tree, occurring in other parts of the picture, is so easily distinguished that we need say no more concerning it. We may, however, observe, that this species is often represented in antiques; also that it was very common in Upper Egypt, since Girgé observed it in Nubia, at Thebes, and especially near Elephantina.

The tree above the lioness very much resembles a tamarind; as is also true of the one which is near to the gigantic serpent. Upon the right of the mountain may be seen the large Euphorbium, whilst near the top of it there is an Acacia growing, the tree which stands in front of the animal called in the mosaic the Onocentaur. There is also to be seen a great thicket of reeds near the building intended to represent the Nile. Other plants also spring up at the side of the thicket, round which crocodiles and hippopotamuses are swimming. These plants appear to be the millet, which, according to Diodorus Siculus, the Ethiopians much cultivated in many of the islets of the Nile (lib. i. p. 24). Finally, under the thicket, and all round it, appear, on the surface of the water, many flowers of the lotus, some of which are blue, and others red. Athenæus has long ago distinguished these varieties (Deipnos, lib. xv. p. 677), and the French expedition to Egypt has made us acquainted with others which had escaped the attention of ancient authors.

Besides this, we shall add, that other rare animals are likewise figured in the mosaics that are copied in the work of Jean Cimpini, entitled *Vetera monumenta, in quibus præcipue musiva opera, sacrarum profanarumque ædium structura,* &c., and printed at Rome in 1790. The mosaic of Palestrina is there represented, but in a manner far from accurate. In plate xxx. may be seen a bustard (*Otis tarda,* Lin.), and also a fish of the trigla or mullet kind. In plate xxxiv. of the same work there is a mosaic, on which we observe the common and the large lobster, also a turtle-dove, a Guinea fowl, and the variegated helix shell of our gardens.

RECAPITULATION.

In reviewing the facts which have now been dwelt upon, it seems clear that many species of the terrestrial mammalia have disappeared from the surface of the earth within the period of historical record. It is also true that one of those species now lost is found in bogs and estuaries, in which are also discovered species which, up to the present moment, have been considered fossil; and hence, these fossil species must have been extinct at the same epoch as the former. It is thus true of the hyena, the rhinoceros, the elephant, and the hippopotamus, as it is of the Irish elk, which is often associated with them;—these should no longer be considered as fossil, but only as inhumated *, since these last named have ceased to exist posterior to man's creation, and the entrance of the seas into their present receptacles†.

In this memoir we have only enumerated among the lost races of which the ancient monuments have preserved traces, five species of the terrestrial mammalia; but we could easily have increased their number, had we not been most scrupulous in our determinations. Thus, for example, we see in plates lxiv. and lxiii, of the works of Micali, which we have already referred to, one of the carnivora engaged with a leopard in devouring a stag and a bull, which carnivorous animal appears to differ from all races actually living. Nevertheless as this animal has some resemblance to the streaked hyena, we have preferred saying nothing about it, though it may probably be a real being, since these plates exhibit more than half a dozen of specimens of it, all drawn with the same characters.

We might have done the same with a great number of other animals which we find represented upon very many other antiques, which so much the more merit our confidence, that the animals whose traits they represent are pourtrayed with fidelity,

* It seems right to express the differences which exist between such organized bodies as have become extinct cotemporaneous with, or posterior to, the creation of man, and consequently the collecting of the seas in their respective receptacles, and the fossils which have been destroyed apparently before these great events.

† We take this opportunity of correcting a grave error, which occurs, however, in the original Memoir, at page 163 of vol. xvi., where we should read, in line 2d, *inhumated*, in place of *human* varieties.

and sometimes with their real colours. Such, for example, are the peacock, the partridge, the parrots, the ostrich, and the horse, which are to be found in the mosaic discovered among the ruins of Italica, in Spain. The horse is represented as he is found in his wild state, that is to say, with a uniform bright bay coat.

What we have just said of the works of Micali, might also be said of many others, amongst which we shall only quote the *Museum Etruscum* of Gorius, the different works of Augustini and Montfaucon, of Caylus, of Hancarville, of Vaillant, and of Mariette. The treatises respecting the various monuments which have been discovered in Pæstum, Pompeia, and Herculaneum, also deserve to be mentioned in relation to the same point.

Lastly, we may remark, that, according to the opinion of M. Schweighæuser, Greek Professor at Strasburgh, there is a lost species, engraved in a work of Millin (Galerie Mythologique), which we have not in our possession, and which we have not been able to procure. According to this able antiquarian, this animal is not intended, as Millin has supposed, to represent the Trojan horse, but rather a species of goat or antelope, quite different from the known races. This supposition is confirmed by the inscription which is found in close contact with this animal : in fact, whether we read it Αιγσα, or Γιγσι, it ever brings us back to goat, or sort of antelope, for the Greek word Αἰξ signifies goat, and in the German dialects they still employ the word *Gegse* to distinguish a variety of this genus.

If this observation be correct, as its author's name would induce us to suppose, it would hence result, that the terrestrial mammalia which have disappeared from off the surface of the earth since the times of history, and of which antiques have preserved the recollection, would belong to the same families as those species which are buried in the quaternary deposits, linked to the same epoch, and which are extinct the same way as are the former. It is, in truth, only with the Pachyderma, the Solidungula, and the Ruminantia, that the lost races brought under our notice in historical monuments connect themselves, and it is known that these are also the families which abound most in all quaternary formations whatever.

To recapitulate, the lost species of the terrestrial mammalia, the traces of which are preserved on the antiques, are reduced, according to our observations, to the number of five, and if the opinion of M. Schweighauser be adopted, they will mount up to six.

These six species may be distributed so as to present, 1*st*, Two pachyderma, the native country of one of which, viz. that known by the name of Ζιθιτ, according to Kircher, is known ; 2*d*, A variety of the Solidungula, intermediate between the dzhiggtai and the quagga; and 3*d*, Two Ruminantia, the one of which is the great Irish elk, and the other the nabum of the Ethiopians, designated in the mosaic of Palestrina under the name of Ταβους ; 4*th*, The last would be the goat or antelope, figured in the Galerie Mythologique of Millin.

Nor are the terrestrial mammalia the only animals of which certain species have been lost since the times of historical record ; for we know that M. Geoffroy St Hilaire discovered in the catacombs of Egypt two races of crocodile, which have not been found elsewhere, and which at present appear lost. It may be said, without doubt, that it is of these races, as it is with the crocodile, the snout of which is furnished with a kind of horn, which Ælian had described, and which he said he saw in the Ganges. For a long time it was regarded fabulous, and the more so because the specimens which were found in the Ganges did not present that horn which Ælian had given them. Within these few years, however, Messieurs Diard and Duvaucelle have afresh discovered this horned crocodile of Ælian, and which the individuals that had been previously seen had accidentally wanted.

And if there are thus reptiles, of which species have been lost within the period of historical record, still more must there be birds and fishes which have become extinct since the same epoch. Regarding fishes, there are a great number of varieties described by naturalists, for example, by Opian, of which we know nothing. In particular, we are ignorant of his *anthias*, which was used to catch the fish called barble, and which consequently was of very small dimensions. This fish is not therefore, as has been long supposed, the red-dory fish of the Mediterranean. As to birds, there is an equally great number of them depicted on antique monuments ; and as the greater part appear to have been designed after nature, we can already state, that, among these species, there are many which have been lost. This, then, is a branch of the subject to which we may return at another time, if the investigation is not undertaken by some naturalist more favourably situated than we are. In truth, all that we have said on the subject of lost races ought only to be considered as a sketch of a work which will doubtless be finished by those who, having the use of grand museums, will thus have under their eyes the originals of those objects of which we have only seen more or less faithful copies.

The facts to which we have just been alluding certainly enable us to judge how many causes there are which unite in producing the loss and annihilation of a great number of the wild animals. We see that, besides natural causes; politics, religion, and even honour, engaging the grandees of Rome to vie with one another in the amusements of the Circus, also co-operated to the same effect. Though less active and less powerful than natural causes, these others have certainly exercised a very considerable influence upon the disappearance of certain animals; and the more so because that sacrifices, and the games of the Circus, and triumphant feasts, led to the slaughter of vast multitudes. To these causes are to be added those which, at a later period, have resulted from the benefits brought about by civilization, which by culture have cleared away forests, and moors, and fens, and which consequently have destroyed the tribes that dwelt there, where they, at the same time, found an asylum and shelter. Thus the destruction of numerous species of wild animals may very well have been produced by very simple causes, altogether in the natural course of events; and to explain these great changes, it is not at all necessary to have recourse to inundations or to violent and terrible revolutions.

Nor, finally, let it be forgotten, that the living races must have had a great tendency to decrease and disappear whenever their mortality was greater than their reproduction. This circumstance must certainly have occurred whenever the individuals of the same race, whether through the influence of man, or by any other cause, were so widely separated from each other, that they could not congregate together so as to breed. This cause, joined to those we have formerly enumerated, has very probably produced the loss of many different races, of which now-a-days we find traces only in the bowels of the earth, or in the writings of the ancients, or finally in those antiques which we owe to artists of a former day.

Enabling Crypto-research – Creating Of Books and Beasts By Matt Bille

Cryptozoology, like any science or aspiring science, is build on a base of information and a scaffold of analysis, if the base is weak, it collapses. so where does the information come from in cryptozoology? Since the topic features prominently on worldwide social media, not to mention television, movies, and publications, the amount of information available—good, exaggerated, or invented—is staggering. Where does the new cryptozoologist, or the cryptozoologist wanting to delve deep into a new topic, start looking? Based on my 45 years of reading, writing, and research, I would say "books."

Books matter because this highly controversial field is barely 60 years old. Much information recorded before that, including the information in letters, diaries, telegrams, etc., is gone. While some have been preserved in archives kept by CFZ, Fortean Times, the International Cryptozoology Museum, and so on, few people can browse those. There aren't many scientific papers or journals. So, it's books. Being focused and relatively permanent, books allow the reader to make a useful selection and evaluate the sources the author used as well as their approach to the data.

There are a several online collections of cryptozoology books, some annotated and some not, but no book collecting reviews and suggesting what books carried mainly zoological information and which focused on the paranormal or presented mainly unfounded speculation. (I'm a science writer: paranormal topics can be intriguing, but they're not zoology.) So, I wrote *Of Books and Beasts: A Cryptozoologist's Library*.

While the great naturalist Louis Agassiz wrote, "Study nature, not books," the regions where new animal species might exist are inaccessible to most people in the industrialized world. Serious cryptozoological researchers who want to understand the field must start with reading.

Part of the value of books is based on the simple fact that they take time to write. Yes, the interested high school student can cut and paste and upload a "book" in days or hours, but for serious writers, just the act of writing books offers some time for reflection. Furthermore, books from the pre-digital age had to pass the eyes of an editor, fact-checker, and sometimes peer reviewers, whose attentions at least sometimes weeded out nonsensical works. Many shoddy books were still published, of course, but I think the system had some effect on overall quality.

In assembling this library, the first challenge was to keep the books to a reasonable number. (In cryptozoology, it seems, the plural of "anecdote" is "book.") I cut out books not written in English, limited the collection to two books I had personally read, and created some other rules to cut down the pile to something I could manage.

I started out in early 2021 thinking this would be a relatively easy endeavor. I'd look over my past book reviews, read a few new books, and kick something out within six months. Ha. I quickly realized that, to provide a decent sampling, with emphasis on more reliable information, was a much bigger project than that. It stretched out to 18 months, in which I reviewed all my past work and searched out and obtained over 100 new books. I also thought that this would be an easy way to get into self-publishing, but as the number of books doubled to the final figure of 400 and my self-imposed deadlines zoomed by, I opted to do two things. The first was to select Hangar 1 as a publisher. The second was to add a scientific editor, the life sciences historian Dr. Anne Larsen, so I could cram in more books while she curated what I had. Both moves turned out to be good ones.

Even the most dedicated researcher can't possibly read all the new cryptozoology books that come out. A multivolume set covering all the books or even all the significant ones would take a team of researchers years to compile and would be immediately obsolete. So, beginning with my back catalogue of Amazon reviews, newsletter items, reviews in my blog, and so on, I tried to ferret out what was most important in the vast sea of information I'd passed over. I read all of Loren Coleman's annual Top 10 Best Cryptozoology Book lists (he created a new category to work Of Books and Beasts into 2021!) and lists by other people and organizations. I asked for recommendations from cryptozoology groups, studied reviews on Amazon and many other sites, made passes though eBay, and wandered the bookstore aisles. I ended with a list of books I would have liked to add but couldn't get to, but such is the lot of the lone researcher.

One important thing I learned is that cryptozoology books are almost entirely written by authors of European descent. It was hard to find books written in English for people who live in areas like South America and Africa, which are after all the regions in which large cryptozoological animals are most likely to be found. I made that a point of emphasis, finding some important books including Monsters of Patagonia and Beyond the Secret Elephants.

I created one room in our library for the cryptozoology books themselves I divided these into a basic library, primates land other land animals, water cryptids, and others, which covered the whole field I cut way down on Sasquatch and Nessie books because they'd overwhelm the library. If Sasquatch is real. I'm willing to bet the number of books exceeds the number of living Sasquatch.

The identification of books as cryptozoological is not quite as clear-cut as one might think. I ended up putting some books in here by Alan Rabinowitz and Forrest Galante, for example, which use the " c word" only in passing but offer important stories of the finding of new and "lost" species.

Something lacking in many cryptozoological endeavors is an understanding of the science underlying any search for new animals. Some cryptozoologists are experts in particular areas of science, or they make themselves so, but many jump straight into cryptozoology and the lack of foundational knowledge of the sciences is usually obvious. While reading alone won't make anyone, say, a primatologist, reading enables the researcher to understand the lingo of the field, ask intelligent questions of the degreed scientists they may talk to, and provide a frame of reference for reading or discarding cryptozoological books and claims.

So, the second room of my library holds a collection of scientific works. I set aside textbooks to focus on books that are authoritative but understandable and available to a broad audience. I included books on the natural world, many types of animals, general biology, evolution, marine biology, and so on. There are classics here as well as recent works. I included some essay collection books from people like Stephen J. Gould and Loren Eisley (the first scientist to ever see the Wyoming mini-mummy "Pedro" and declare it an abnormal infant).

The third room houses the fiction, which of course was a lot of fun to read and review. I think fiction has a place in any such reference collection, not only because it's enormously influential in shaping the public's view of cryptids, but because cryptozoology is often short on agreed-upon facts and it's useful to read authors' speculations on why sasquatch lives in a particular place or why a sea serpent has gone uncaptured. (There aren't a lot of sea serpent books, though.)

Crypto-fiction is a broad slice of literature that includes many genres. I set aside most "human-eating monster" horror, however entertaining a book titled *Swamp Monster Massacre* may be, and sampled classics like *The Lost World*, Jaws (not an unknown species, but a major trend-setter of a novel) and recent efforts like *Devolution*, a best-seller that will not be popular among the "loving forest people" sasquatch enthusiasts. I also left out magic or human-created creatures, although *Jurassic Park* demanded and received an exemption. I bent the "book-length" criteria to slip in Joe Hill's story "By the Shores of Lake Champlain," a uniquely memorable take on a famous cryptid. Every library has (or should have) an "Oddities" shelf, so I created a final section that included famous but less useful cryptozoological works, a few extra Bigfoot books, and so on. I included fiction that violated some criteria for the main fiction section:

Jim's Butcher's *Working for Bigfoot*, a tale of his wizard Harry Dresden, is a hoot, and Jonathan Maberry's genetic-engineering tale *The Dragon Factory* includes the vital information that dodo does not taskset like chicken). I also included some books on woodcraft I asked cryptozoologists to recommend, since I'm worse than useless in the field.

This is "A" Cryptozoologist's Library and not "The" Cryptozoologist's Library. This work began with my collection, which inevitably means it's heavily influenced by my own curiosity. While book reviews, by definition, are opinions, I tried to read everything as a skeptic in the proper sense of the word: someone who wants to see new scientific discoveries but needs convincing evidence.

While I've written two very well-received books on zoology and cryptozoology, I did not set out to make an authoritative list of the best books. I knew every well-read cryptozoologist (and many a cryptozoology author) could point to books they thought should be included, and to critique the content of my reviews. I think a second edition should come out in five years or so. Until then, happy reading!

THE ANIMALS THAT SHOULD EXIST

Lars Thomas

There was a clearing in New Zealand, where it used to fly.
There was a clearing, where its loud warbling call used to wind through the cool morning mist like dancing fairies in a mysterious bog.
There was a clearing, where it used to fly, the huia-bird with its strange sexually dimorph beak and its long white-edged tail.
There was a clearing, where the huia-bird was hunted for it's beatiful tail-feathers.
There was a clearing – but not anymore.

The observation tower was very tall, and the clearing was very large. So was Pureora Forest on the North Island of New Zealand. It had taken me at least a couple of hours walk to get there. It was late spring, early summer, and the air was filled with birdsong, albeit mostly the sound of European species tingling among the trees. So many species have been introduced from "home" that they are drowning out New Zealands native species. It might as well have been a sunny May morning in Northwest Europe.

But then again…

When I reached the observation platform, high enough to be able to look down on the tops of the trees growing in clusters dotted all over the big clearing, a bird the size of a jackdaw took off from a branch almost directly below me, and flew off. It didn't make a sound, and all I saw for a few seconds was its back and tail, before it disappeared among the trees further off. But oh, what a tail – it was black as night, and the tips of the tailfeathers looked like they had been dipped in white paint. Not particularly spectacular some would say, and they would be right, but the only bird ever to fly around in a New Zealand forest with a tail like that, was the huia, and that early summers day in 1991, it was something like 85 years since the last confirmed sighting.

The huia (*Heteralocha acutirostris*) was, or is possibly, in many ways a unique bird. First and foremost because of its very special bill. The male had a powerful chisel-

© 2009 Encyclopædia Britannica, Inc.

shaped bill that would have been a credit to any self-respecting woodpecker, whereas the bill of the female was long, thin and curved. Working together they could break out insect larvae from the deepest cavities in tree-trunks – and they would split the

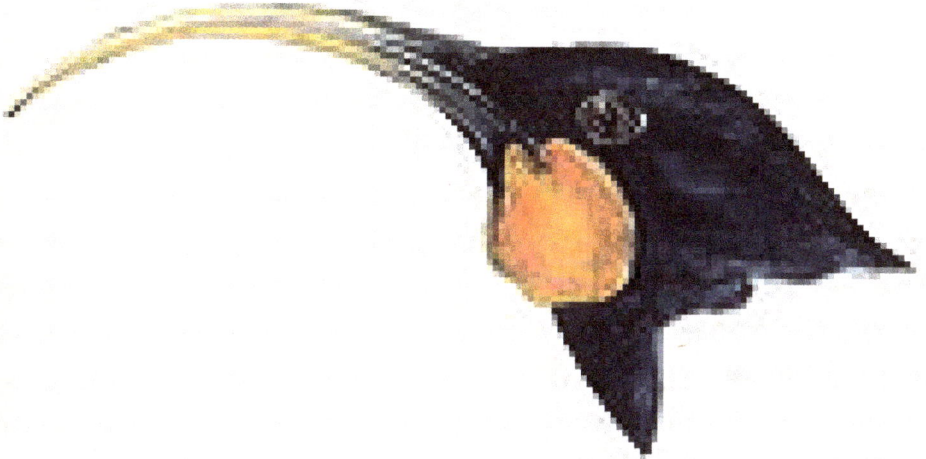

catch. They thrived in the forests of the North Island, but like so many other island-living birds, they ware far too trusting towards humans. And that, as history have shown numerous times, is a tactical error of some magnitude.

They were shot, they fell as prey to various imported predators, and most important of all, their forests were cut down, and their clearings ended up as fields. But they were also the victims of scientific curiosity and collector's mania, when the museums in their eagerness to get specimens of the strange birds for their collections, forgot to think ahead. It was the exact same attitude, when just a few decades earlier in 1844, the great auk was pushed over the edge from a small island of the coast of Iceland.

In just a single month in 1886, 646 huia-skins were collected in the southern Hawkes Bay area, and sent to various museums around the world. This alone took a serious toll of a bird that was probably not that common to start with. And the tragic thing is of course, that there was no need for it to end like that. The huia should still be able to exist today. Its two nearest relatives, the saddleback (*Philesturnus carunculatus*) and the kokako (*Calleas cinereus*) are still with us. Rare? Yes! And with a lot of help from humans? Yes! But at least they are still here!

So, did I actually see an "extinct" bird? I still don't know more than 30 years later, but I do know, I looked for the thing for hours without finding it again. Although it should have been there somewhere. But then again, that's the thing with so many different animals. Some of them shouldn't exists at all, but seem to do so anyway, and some of them don't seem to exist, even though they should.

Turn the picture upside-down
I have spent large parts of my life and career studying animals that exists with one or perhaps (maybe) a couple of legs in reality, and the rest of the legs in some form of twilight zone, a grey and misty border between being and not-being.

It is possible to study a large selection of living organisms in a very hands-on sort of way – bodies, skins, scales and other physical remains can be found in abundance in the drawers and cabinets of museums, and should you so wish, it is possible, albeit sometimes with a certain amount of difficulty, to go out into the wild, and study most af these creatures in a living state. By and large, this is the territory where classical zoologists can be found. I have been there myself, at first in the form of an eager young birdwatcher, later as a professionally trained marine biologist, and in later years as a field entomologist.

In between all that, I have also worked as a cryptozoologist, i.e. someone who studies hidden animals (from the Greek word kryptos = hidden). This is a world of animals and strange creatures where the word "should" crops up on a regular basis – usually in the form of "should not exist". As a cryptozoologist you study animals, or rather most of the time accounts of animals, living outside the searchlights of classical zoology. Animals living in places where they shouldn't be. Animals living (or at least being seen by people) at times where they shouldn't exists (i.e. officially declared extinct). And finally, of course, the favourite prey of the cryptozoologists; animals that shouldn't exist at all, animals taken directly from the pages of a book of fairytales and folklore – dragons, thunderbirds, seamonsters and a plentitude of all things large, dangerous and scary.

But suppose – just suppose – we turn it all over? Turn the picture upside down, holds it at arms length and squint a little bit? What about looking at the animals that should exist?

Guilt, shame, and a glimmer of hope

We could of course start by looking at a group of animals that are extremely different and essentially not especially closely related, but woven together in a common network of disappointment, anger, guilt, shame and a tiny ray of hope. All we know about them is the fact that they used to be here, but not anymore. They still exist – in a way – in the shape of memories, stories, pictures, skins and museum specimens, but they are not alive anymore. Everyone of them can tell their own story of falling numbers, fewer and fewer places to live, increasing pressure from hunters and collectors, and diminishing hope.

The most frustrating thing is, of course, that it ought not to have ended in this way. There could so easily still have been room for them today. They should still exist. But basically we only have ourselves to thank. We should be ashamed of ourselves, and to some extent maybe we are. This is probably why researchers and foundations around the world have started projects to recreate extinct animals by cloning, DNA-techniques and advanced breeding.

But is this actually a good idea? Just as one can claim, that the animals should exist still, one can also ask whether we should actually be bringing them back? Can we be certain, that the hole their disappearance left in nature, hasn't already closed, leaving no place for them? It would probably be good for our feelings of guilt and shame, but would it also be good for nature? One day we may have a recreated passenger pigeon or a mammoth, but then what? Are they to be released back in the wild? What happens if their presence starts pushing other species out? What happens if things for some reason go terribly wrong? Do we remove or perhaps exterminate the new species we have only just "put on the market"? If history tells us anything, it's the fact, that it can go terribly wrong when humans starts to meddle in the affairs of nature without a lot of thought. The rabbit and the cane toad in Australia spring to mind.

And what about the legal stuff by the way? Is it possible to patent a new species? Someone is bound to try. And will they actually be "thylacines", "mammoths" or "passenger pigeons", or just look-alikes?

Another important issue to consider is the way humans will be regarding the new animals. If it becomes possible to recreate extinct animals species from skins, tissue-samples, teeth or other form of physical remains, why use time and energy to preserve the animals and their environment while their are still alive? Why not just "build" a set of replacements?

Is it in fact guilt and shame, that makes us want to bring the animals back from the death? And is it perhaps a sprinkling of hope that makes us claim we still see them, even if their are not here anymore? Just about every single extinct animal has been seen again and again, years or even millenia after their supposed demise. Our world has so many isolated spots for them to hide in, and you never know...

Rare and strange animals have a tendency to hide themselves in remote locations, the deepest of forests and the tallest of mountains. In the Middle Ages nobody doubted the existence of griffons for example, but South European and Arabic scolars were convinced they were to be found in the mountains of Hyperboria (The Nordic countries), whereas people in the Nordic countries were equally convinced griffins were to be found around the Mediterranean or in the Arabic countries.

What about the great auk, to take one example out of many? It should still be here. All the other species it was living amongst 200 or 300 years ago, razorbills, guillemots, puffins and so on, are still here. Officially the great auk died out in 1844 when the last three documented individuals were killed on a small island off the coast of Iceland. But all through the 1800's there were sightings of great auks along the coasts of Newfoundland or several locations in the Davis Strait between Greenland

Great Auk

and Canada. In the summer of 1984 I spoke to several old locals who had lived on the small archipelago of Kronprinsens Ejlande in Disko Bay, and they insisted that great auks were shot in the area occasionally back then. One of them even remembered plucking one, when she was a little girl.

Studying phenomena like that, is like walking out into a clearing at the end of a long walk in a deep dark forest. First you get a few glimpses of the sun, when the first sightings start to trickle in, but not many, and they are few and far between. But suddenly there is some sort of breakthrough, the animal, the phenomenon is lodged in people's consciousness, and suddenly everybody sees the great auk or whatever we are dealing with, and we step out into the a clearing filled with blazing light and hope. When everybody sees it, there must be some truth to it, and in just a minute or a day, we are going to find it, and all will be well. But nothing happens, the number of sightings start to drop, the images start to fade, and we all walk on into the woods, leaving the clearing of hope behind us.

Being lodged in people's consciousness is a phenomenon that I think plays a far bigger role when it comes to sightings of animals that should exist. The thylacine is a excellent example. This rather weird looking dog/fox/wolf-like marsupial with a striped behind is the archetype of an no-not-extinct-after-all animal. It's an expert in showing itself in short tantalizing glimpses in small clearings in the depths of the Tasmanian woods – or scaring the pants of late-night drivers.

The last documented thylacine died in Hobart Zoo in 1936, because somebody forgot to let it into its sleeping cage on what turned out to be one of the coldest nights in a very long time. Next morning the poor thing had died from exposure.

The thylacine should still exist – nothing really important has changed in the Tasmanian landscape since then, and judging from the number of people claiming to have seen living thylacines since 1936, it very much does. We are not talking a few, or even a couple of dozen sightings, we are talking thousands of sightings. The large number of sightings have of course convinced many people that "there must be something to it". I have talked to several people returning from a trip to Tasmania more convinced than ever, that the animal still exists.

But I have also been to Tasmania several times, interviewing witnesses and studied the lay-out of the land, so to speak, and I have come back with the exact opposite feeling, i.e. less convinced. I think the extraordinary number of sightings actually indicates that there is nothing out there (I would love to be proved wrong, though). So many sightings without a shred of physical evidence can only mean, that the only place the thylacine lives, is in the consiousness of people. It's incredibly difficult to avoid being at the receiving end of a bombardment of thylacines when you visit Tasmania. The animal is, if not on everybodys lips, then at least on everybodys luchboxes and drinking containers. You find them as stuffed toys and in Saturday morning cartoons. Thylacine art? All over the place, thylacine posters – you bet! On bus stops – indeed! Fridge magnets, pens and notebooks? But of course. It's even the focal point of 2011 movie "The Hunter" with Willem Dafoe. It takes up so much space in the minds of the people on the island, that every time they see something they can't identify straight away, the thylacine takes up so much space in the memory -file marked "Possible Explanations" that it's always the number one pick.

Where are the others?
In the storerooms, drawers and cabinets of the museums of the world, you will find innumerable animals only known from that one single specimen. But where are the

others? Why have we only found one? Have they gone extinct since then? Did the zoologists actually find the last individual? What has happened? They should be out there.

In 1883 Danish naturalist William Sørensen found a fairly large spider (15 millimeters in bodylength) at the foot of a heather-clad hill in an area called Mols Bjerge in Denmark. He described the spider and gave it the scientific name *Lycosa danica* (this has later been changed to *Pardosa danica* should you care to look it up). The new species was placed in the collections of the natural history museum in Denmark, and all was well – except for the small fact, that it has never been seen since! This is especially strange because Mols Bjerge was later turned into a natural history field laboratory, and it is probably the most thoroughly studied piece of land in Denmark. But no, not a trace of the thing. And there has actually been several dedicated searches for it.

You can of course always speculate on the reasons for the loneliness of *Pardosa danica.* It was already dying out at the time, and the naturalist accidently stumbled on one of the few remaining animals. It's actually an entirely different species normally found in another part of the world, and the one found in Denmark was a single specimens blown way of course. It's a malformed/mutated specimen of a another more common species, or it is a simple administrative error, a miswritten label or some sort of lack of information making the animal more strange than it actually is.

Wrong information and administrative errors or sloppiness is the nightmare of every museum (after fire and vermin in the collections). Bad information means the specimens loose their scientific value, or you won't know where to look for them in nature, should you so wish.

Take for instance Delcourt's giant gecko (*Hoplodactylus delcourti*), which we know should exist, since the Natural History Museum of Marseille are the proud owners of the world's only (albeit extremely worn and battered) known specimen. And here we are talking about an animal which should be reasonably easy to locate seing it's more than 60 cm in lenght, almost twice the size of the world's second largest gecko.

The giant gecko was described in 1986 based on a mounted specimen which had been in the posession of the museum for more than 100 years without anybody giving it a second thought. And then a couple of herpetologists started on a sort of spring cleaning and realised something was afoot. Unfortunately there was hardly any information about the specimen in the files of the museum apart from the name of the man who had collected the animal. This has made it more than a little bit difficult to establish where to look for additional specimens. The animal was probably collected sometime between 1833 and 1869, and it was found somewhere in the southwest Pacific area. The most likely place being New Zealand, because its resemblance to the kawekaweau, a kind of forest demon from Maori legends. This legendary animal is supposedly more than 60 cm in length, thick as a man's wrist, grey or brown, and a terrifying omen of death. There is in fact a tantalizing note claiming, that the James Cook expedition to the area in 1777 did catch a couple of large reptiles, but they never made it back to England.

If the giant geckos were forest animals, why are they not here anymore? There is still plenty of forest in New Zealand, North Island and South Island, although not necessarily as much as before. They should still exist. On my second trip to New Zealand in 1993 I was out looking for sleeping owls in a forest on the North Island. A local birdwatcher had pointed me in the direction of a group of very tall trees often used as an owl roost. I never found the birds, but at one stage it looked to me like a large piece of bark broke off one of the trees (something like 20-30 meters up), but instead of following the call of gravity, it glided upwards and disappeared around the back of the tree trunk. I am not claiming anything, but tree-living geckos often look like bark.

When does an animal exist?
In the world of zoology an animal doesn't exist before it has received a scientific name, has been properly described, and a reference specimen, a so-called type, had been deposited in a museum. This makes it extra interesting that we also have a selection of animals whose physical existence can in no way be doubted, some of them are known to exist in their thousands, but officially they don't exist at all. We don't know who or what they are, or for that matter where to find them. If you want to try your hand at describing a new species, there are excellent opportunities if you

delve into the world of exotic pets and fishes. There is quite a number of colourful fishes, bugs and beetles without a formal scientific description even though they have been part of the exotic pet trade for years, even decades.

You can of course also just walk outside to find an example of this. The harvestman *Leioburnum sp.* A, first found in Europe in Holland in 2004, has been extensively studied. Since its discovery in Holland it has spread to a good deal of Northern and Central Europe. In Denmark it has only been found in a handful of places, and the only apparantly viable population lives in a handful of holes in a single piece of wall in the castle ruins in the town of Vordingborg. But it still lacks a scientific description, It may be a completely new species, it may have come here on a ship from a part of the world where it is common. But nobody has bothered to connect the dots. It exists, right here, right now.

And yet, here we are, right here, right now, watching hundreds of species slipping through our fingers, like dodoes waiting at the exit. And in most cases we have no idea what is happening. We don't know enough, people don't know enough. Should they not…? Butterflies disappear. Their clearings end up as fields, feeding grounds for cattle. They may even just disappear – should they? The learned debates. Should they be here? There should be room for them. The birds disappear. The fishes disappear, the insects disappear – they should be splattering on front windows everywhere. People lose a sense of the big picture – "there are no butterflies in my garden this year" – "What's happening?" – Shouldn't they…? "There are lots of butterflies in my garden this year." – "What's happening?" – Shouldn't they…? "We haven't seen a single starling this year." They at least should…

What do they actually mean, those two tiny words- "should exist"?

There is another clearing in New Zealand, where it should exist.
There is another clearing with an enormous tower, where you can see all that should exist.
There is a clearing, where I went up the stairs to the tower an early summer's day in 1991.
There is a clearing, where a black bird flew directly away and disappeared, when I looked.
There is a clearing, where a black bird with a white stripe on its tail flew off, not to be found again.
There is a clearing, where only the huia-bird has had a tail like that.
There is a clearing, where it should exist.

Index to Flying Snake issues 6 to 10
Richard Muirhead

(Continues from index in 2020 CFZ Yearbook)

Flying Snake

A Journal of
Cryptozoology, Folklore and Forteana
December 2020   Vol 6 Number 18   £3.99

Elephant Shrew Rediscovered . New species of Owl in Trinidad . Meet The Centaur . Albino Hedgehog . Forteana from Fate . Genets in the UK . Singing dogs rediscovered . The Metallica Snake . Weird Jumping Porcupines

Monster crab,Suffolk,1857   Mar-16
Mouse,peculiar,Inverness,1834   Dec-14
Mystery big bird Apr-14
Mystery birds,Roman   Apr-14
Octopi,aggressive   Apr-14
Octopi,mystery   Apr-14
Octopus, W.Virginia,1954   Apr-14
Odd sea creature,U.S.,1896 Dec-14
Petrified cat,1870   Apr-14
Prehistoric bird,Wyoming,1926   Mar-16
Pterodactyl,California,1908 Apr-14
Pygmy horses,India   Apr-14
Python,Florida   Dec-14
Rabbit,huge   Sep-15
Rhinoceros,undescribed   Sep-15
River monster,Iowa,1885   Sep-15
Romans in Ireland   Nov-13
Ropen,Australia Sep-15
Saints Paul & Peter in England   Dec-14
Seal or dugong? Dec-14
Seal,pied,Chester,1766 Sep-15
Seals,S.China Sea   Sep-15
Sea monster,Cornwall,1786   Apr-14
Sea monster,Macau   Mar-16
Sea serpent,African   Mar-16
Serpent,huge,Grenoble,1809 Apr-14
Sea serpent,flying,1911   Sep-15
Serpent or eel,France,1948   Apr-14
Shark in Australian river   Sep-15
Siam,cryptozoology   Dec-14
Sign from heaven   Mar-16
Snake,feathered   Mar-16
Snake,huge,Australia,1824   Mar-16
Snake,winged,Australia,1937   Mar-16
Snake with legs,Africa,1950 Sep-15
Snake with legs,Texas,1899 Sep-15
Snakes eggs in stalagmite   Dec-14
Snakes,thousands   Mar-16
Spider,giant,Bristol,1714   Apr-14
Spiders,giant,Colorado Nov-13
Spider,giant,London,1798   Sep-15

https://commons.wikimedia.org/wiki/-
File:Cressie_at_Roberts_Arm_Newfoundland_2016_by_Dale_Gilbert_Jarvis.jpg

The Monster of Crescent Lake, Newfoundland
David Weatherly

"Welcome to Robert's Arm. Beothuk Trail, Lake Monster Country. The Loch Ness of Newfoundland." proclaims a plaque near the shore of Crescent Lake in northern Newfoundland. Near the sign is a statue of a dragon-like, serpentine monster with scales and vicious teeth.

The statue grabs one's attention, and for those unaware of the area's monster lore, it piques the curiosity. Of course, the statue of the beast is far from an accurate portrayal of the lake's mysterious creature. The lake's monster, known as Cressie in a nod to the famous Nessie of Loch Ness, Scotland, is described as a long, serpent like beast, dark brown to black in color and resembling a giant eel. Frequently, connections are drawn between Cressie and native tales that mention creatures known as "woodhum haoot" or "pond demon," and the "haoot tuwedyee" or "swimming demon."

Whether or not there are specific native stories about a monster in Crescent Lake is debatable, and, at the least, no firm accounts from natives have been found to date.

Lore surrounding the monster comes primarily from the community of Robert's Arm. Robert's Arm is a small town built up from an earlier settlement that was established in the 1850s. The town, formerly known as Rabbit's Arm, remained a small community for decades. Even today the population is under a thousand residents. In the early 1930s, logging was the main economic core of the area, and, for a long time, the greater portion of the residents had some family members employed by the timber industry.

According to area folklore, there have long been tales of something monstrous living in the depths of the lake. The first water monster report most frequently cited involves an early resident of the area; a woman known as Grandmother Anthony. Sometime around 1900, the old woman was out picking berries near the lake when she spotted a large, serpent like beast swimming in the water. Grandmother Anthony rushed home and told others her tale. Unfortunately, no other details of the sighting have been passed down.

The next notable encounter wasn't until the 1950s when a pair of local men spotted

the beast. The two men saw an object in the water that they first took to be a stick or log, then an overturned boat. Concerned that someone had capsized and was in trouble, the men jumped into their boat and raced out to see if anyone needed help.

Approaching the object, the men were puzzled to see that it was moving into the wind, rather than being blown in the same direction of the breeze as a boat would be. When they reached the thing's position, they received a shock—the object turned out to be a living creature—black in color and large and rounded in shape. When the boat approached, the thing took its leave, slipping beneath the surface and vanishing.

The men didn't wait around to see if the creature would resurface. According to Andrew Burton, one of the two men involved, the pair wasted no time getting away and rushing back to shore for safety.

On June 7, 1960, four loggers working near Crescent Lake spotted something in the water. Like Burton and his cohort ten years earlier, the loggers initially thought the object was an overturned boat. As they observed the moving mass, they realized it was a living creature. It swam along near the shoreline then crossed over a sandbar without any difficulty.

One of the witnesses, Bruce Anthony, reported that about ten feet (about three meters) of the creature's length was visible when it navigated the sandbar.

By this time, the creature was a local legend and those living around the lake knew the stories of something strange in the water.

An article titled *Monsters of the Maritimes* by John Braddock ran in the January 1968, issue of the *Atlantic Advocate* but otherwise the creature didn't receive any mainstream press. That changed in the 1990s when a string of reports put the beast on the map.

On a spring day in 1990, an unnamed witness spotted a sleek black object swimming in the lake. It was around nine o'clock in the morning and the object was spotted at the lake's southern end. The witness described it as a large animal that rose to about five feet (around 1.5 meters) and created an area of churning water from its movements.

Some people have tried to dismiss this and other sightings, claiming that witnesses are seeing large, floating logs that have come up from the lake's bottom, or perhaps an otter or family of otters swimming across the water. It's important to note that area residents see the lake on a daily basis and are familiar with common objects and animals found in the region. This being the case, locals are adamant that they aren't mistaking common objects or animals for a monster.

There wasn't much of a gap in sighting reports this time. Around noon on July 9, 1991, the creature was observed by a man named Fred Parsons and his wife. Parsons was a retired schoolteacher and newspaper correspondent living in Robert's Arm. He was also a recipient of the Robert's Arm Citizen of the Year Award and as such, he was considered a reliable witness.

On the day of their sighting, the Parsons's were driving near the lake and had a good view of the water. By habit, Fred Parsons kept glancing out at the lake. He was shocked when he spotted an unusual creature, and he stopped the vehicle to get a better look at the animal.

Parsons said the creature was serpent-like and dark brown in color. It moved with an up and down undulating motion like a mammal, and it swam on the surface of the lake. He estimated that the creature was about twenty feet in length (around six meters). In a later interview, Fred Parsons stated that he thought the thing looked like a "giant eel."

The next notable sighting came in the fall. On September 5, 1991, at 4:30 in the afternoon, local resident Pierce Rideout was driving his pickup truck along the lakeshore when he saw movement in the water almost five hundred feet from shore (about 150 meters).

At first, Rideout thought the movement was the bow wave of a small boat, but as he continued to watch through the open window of his truck he realized it was something else—something living. The object vanished below the surface of the water, and in moments it reappeared, pitching forward in a rolling motion.

Rideout said the creature was around fifteen feet (about 4.5 meters) in length and black in color. No fins, tail or fluke were visible, and the man was unable to see the creature's head or neck. The animal suddenly plunged beneath the surface again and was gone. The witness had watched the creature for approximately three minutes.

Rideout, a farmer and part time mailman in the area, had a reputation as a no-nonsense type who spoke his mind. Ironically, he had been a vocal skeptic of the lake's monster stories but seeing the beast for himself set him firmly on the other side of the fence. He was so shaken by the experience that he started carrying a gun with him whenever he had to journey near the lake.

Longtime area resident Effie Colbourne had the longest recorded sighting of the lake's monster, a sighting that lasted a full fifteen minutes according to her report.

Colbourne's sighting occurred while she was at her home on June 6, 1995. She was near a window with a nice view of the lake and was lost in deep thought when movement out of the corner of her eye grabbed her attention. She went to the window and gazed out to see a mysterious creature swimming across the lake.

"I definitely saw a head and a body, whatever it was, it was real," says Colbourne. She described the creature as serpent like and twenty to thirty feet (around 6-9 meters) in length. The head was like that of a horse, but slimmer and more pointed. By the time of her sighting, Colbourne had lived on the lake for decades and, growing up, she'd heard old timers in the area talk about something monstrous in the water.

Reports of Crescent Lake's monster drew international attention during the 1990s. The *Augusta Chronicle* out of Augusta, Georgia, U.S., covered the beast in its March 21, 1999, edition, giving it a whole article under the banner "Cressie of Crescent Lake—A Monster Eel?

In July 2000, Robbie Watkins and Richard Goudie were part of a crew working along the Hazelnut Hiking Trail near the lake when they spotted a creature in the water that they believed was Cressie. Their description of the monster matched previous reports and the body of lore continued to grow.

On August 14, 2003, Canada's *CBC News* ran a story about the lake monster with a report from Robert's Arm resident Vivian Short who said she'd seen the creature.

Short was another local who had initially doubted the existence of anything strange in the lake, but her personal sighting changed her perspective. She said that she was driving with a friend near the lake and when they rounded a bend in the road, they spotted a serpent like beast with a fishlike head. She told *CBC News*:

"We saw Cressie! We saw Cressie!" Short added that the beast was very large in size and that it "could eat four or five people if they were swimmin'"

The *News* reported that other locals had seen the creature as well. Town clerk Ada Rowsell remarked, "I've had several reports of people sighting some kind of huge monster or sea serpent or some kind of a fish."

By the early 2000s, media reports on the lake's legend had attracted the attention of television producers. The Robert's Arm newspaper, the *Nor'wester*, ran a story in its June 18, 2008, edition announcing that Cressie was about to get international attention with a television spot on *History Channel's* popular cryptozoology themed show *Monster Quest*.

The episode, titled *Lake Monsters of the North*, (season 2 episode 13) included interviews with locals as well as an active investigation into the lake's monster. The program's primary focus was on the possibility that Cressie is a giant eel.

The idea that Cressie is a giant eel is the most popular theory to explain the lake's monster. It's been suggested that the creature is either a misplaced ocean dweller, or an undocumented freshwater species that lives in the lake's depths.

According to some reports, people have even seen eels in the lake and by some accounts, the creatures are vicious. In a November 2, 2019, interview with *CBC News*, Nicole Penny, archivist at Memorial University's Folklore and Language Archive, related a story about divers who went into the lake in the mid-1980s. A plane had crashed into Crescent Lake and a law enforcement team went down hoping to locate the body of the pilot. According to Penny:

"Scuba divers who braved the depths of the lake in hopes of finding the pilot found themselves surrounded by a school of vicious eels. Their bodies were said to be about as thick as a man's thigh. The eels proceeded to attack them and luckily the divers were able to retreat without getting too seriously injured."

The dive team was reportedly from the RCMP (Royal Canadian Mounted Police) and the account is one that has circulated for years. There is, however, a problem with the account—the RCMP has no record of the investigation!

Some people believe that the official record was either lost or that the RCMP just won't mention the giant eel encounter. Unfortunately, the account has been repeated so often that there are variations of it floating around. Most often, the date of the purported encounter is different, although in some cases, the divers were reportedly investigating a hole in ice on the lake in the belief that a snowmobile had plunged into the frigid water.

Law enforcement/eel encounters aside, there is at least some evidence to support the theory that there's a species of eel in Crescent Lake.

There are species of eel that live in both salt water and freshwater. The most likely candidate of eel in Crescent Lake is the American freshwater eel commonly found on the eastern coast of the North American continent. The eels spawn at sea and as they mature, make their way inland to freshwater via rivers and streams where they spend their adulthood.

Crescent Lake's only link to the open ocean is Tommy's Arm Brook—about two miles (just over 3 kilometers) distance from the Atlantic. Most people say that no eels have been reported in Tommy's Arm Brook, so if eels aren't traveling that route, how would they be in Crescent Lake?

Writing in his book *In the Domain of Lake Monsters*, John Kirk mentions an interesting anecdote—some long-time residents of Robert's Arm recall periods when their drinking water tasted brackish. This implies the possibility that the depths of Crescent Lake could be sea water.

If this is indeed the case, it lends credence to the potential that a species of giant eel lives in the lake. Eels are bottom dwellers and would be able to spawn in the depths, leading to their continued presence at Crescent Lake.

A thriving eel population would also explain variations in size reported by witnesses. Different eels—different ages—different sizes. However, it's not an open and shut case. As much as it seems to fit, there are some issues with the eel theory.

The largest know species of eel is the European Conger which reaches an average length of five feet (about 1.5 meters). One particularly large specimen was found that measured almost ten feet (about 3 meters) in length.

This is a far cry from the reported proportions of the Crescent Lake beast. In addition, eels tend to be nocturnal creatures—yet most sightings of Cressie take place during daylight hours.

Beyond this, eels are bottom dwellers so seeing them swimming across the lake's surface would, at the least, be most unusual.

So, what is Cressie? Obviously, the jury is still out, and the creature's identity remains a mystery. Perhaps as more sightings are recorded and as technology continues to develop, we will have a better picture of what lives in the depths of Crescent Lake. In the meantime, Cressie has become thoroughly ingrained in local legend and lore. The monster statue and sign that greets visitors to the area is a reflection of the local attitude that the stories will attract tourists' attention and dollars, and encourage people to explore the area's natural beauty.

It's a great place to enjoy a scenic lake, a magnificent hiking trail—the Hazelnut Adventure Trail—and maybe, just maybe, catch a glimpse of a mysterious lake monster.

(Pic: ERIKA GROENEVELD)

The Van Meter Creatures
Shane Lea and Richard Muirhead

Towards the end of September and in early October 1903, a group of winged pterosaur-like creatures frightened residents of the town of Van Meter, Iowa, U.S.A. Perhaps the most surprising thing about these cryptids apart from the fact that they may have been living pterosaurs , is the fact that they are so neglected by published books and web sites within the discipline of cryptozoology. For example, a search on Dr Karl Shuker`s Shuker Nature blog using the words "Van Meter" scored no hits. We could not find Van Meter in Ronan Coghlan`s comprehensive Dictionary of Cryptozoology, Michael Newton`s Encyclopaedia of Cryptozoology which gives a long summary of living pterosaurs from before the Twentieth Century onwards. George Eberhart`s two volume Mysterious Creatures also does not mention it. I did ask Dr Shuker what he thought the Van Meter cryptids were but as of late November 2021 I (Richard Muirhead) had not received a reply. Fortunately there is enough information on these cryptids online to summarize the story in some depth.This is how the Cryptidz fandom web site summarizes the story: https://cryptidz.fandom.com/wiki/Van_Meter_Visitor

"The strange events occurred in October of 1903. Several respected members of the community told of a mysterious half-animal, half-human winged creature that terrorized some of the town's residents during several nights in the course of the week. Descriptions of the beast suggested that it had large bat-like wings, left a terrible stench wherever it went, and, even stranger, it fired beams of bright light from it's forehead. The bizarre account recalls how several of the locals attempted to shoot the beast but their gunfire didn't appear to have any effect. Fed up with the menace, a group of townsfolk banded together one evening and pursued the creature to an abandoned coal mine. There they confronted not one, but two of the beasts, which both turned and disappeared down into the gloom of the mine…"

It is our opinion that there is no reason to suppose that the Van Meter creatures were some kind of paranormal entities. It is the opinion of Chad Lewis, Kevin Lee and Noah Voss, authors of `The Van Meter Visitor A True and Mysterious

Encounter with the Unknown` that the creatures were paranormal. Indeed, the Van Meter cryptids have all the hallmarks of the Ropen, the pterosaur-like cryptid of Papua New Guinea and elsewhere in the world, including the U.S.A. The flying snake of Namibia (see the Centre for Fortean Zoology Yearbook 1996) of 1942 also had a light on its head. The paranormalists point to the fact that when shot at , the bullets seemed to have no effect. However, an easy explanation for this is that the shots may have missed, or -the creature's skin may include a substance or covering that makes bullets ineffective. As Shane wrote to me on November 12th 2021: " Logically, there are two explanations for the Van Meter Cryptid. It is either a known zoological animal like a Great Blue Heron—4ft tall ,with a 6ft wing span, or some type of unknown flying reptile, possibly a living fossil. Either way,you are talking about a natural explanation for the phenomenon." The Great Blue Heron is distributed across North America. The facts surrounding the appearance and disappearance of

Image from Van Meter Visitor web site https://cryptidz.fandom.com/wiki/
Van_Meter_Visitor

the Van Meter creatures are as follows: For several nights one or more huge pterosaur-like creatures with lights on their heads emerged from an old abandoned mine near the town of Van Meter. Several respected townsfolk witnessed the man-sized, bat-like creatures with a light on their horns moving at speeds "the townsfolk had never witnessed before" (The unsolved mystery of the Van Meter Visitor dailymail.co.uk) They also emitted a foul stench, rather like the flying snake of Namibia.

On the first night they flew above the roof tops. The second evening they were spotted by the town doctor and bank cashier Peter Dunn who took a plaster cast of one of their large three toed tracks (also a possible feature of a flying reptile). The cryptid was seen with a juvenile in the mine shaft, down which they descended, accompanied by a barrage of shots from a posse of the town`s men folk.

On September 24th, 2021, I posted a question on the Living Pterosaurs of the World Facebook Group, asking if the Van Meter cryptid could have been a Ropen-like creature. J.S.G. replied: "It absolutely was, but what`s intriguing about this case is that the creature was able to admit light." Jonathan

Source of image: havhttps://vignette.wikia.nocookie.net/isle/images/2/29/
The_isle_pteranodon_new_2020.jpeg/revision/latest/scale-to-width-down/2000?cb=20200825030749

Whitcomb in his book `Searching for Ropens And Finding God` reports pterosaur-like animals in Kansas which neighbours Iowa. He also reports a sighting of a Ropen in Minnesota to the north of Iowa in September 2011. On January 2nd, 2013, an U.S. Marine based at Fort Leonard Wood, Missouri, saw a Ropen-like creature flying overhead. Missouri is to the south of Iowa. It is interesting to speculate that the small group of Ropens that turned up in Van Meter in 1903 were looking for fish to eat, because the Raccoon River passes through Van Meter and Jonathan Whitcomb has speculated in one of his books on the Ropen that the cryptid exists in coastal areas of Papua New Guinea. The live pterosaurs blog has a report of a pterosaur in southern Minnesota in 1995 reported to have a horn on its head, eating fish (livepterosaurs.com). It was seen at an entrance to a cave. The flying snake (if that is what it was) of Namibia also emerged from a cave. In 2020 Jonathan Whitcomb wrote on the Living Pterosaurs of the World Facebook group: " We have had at times over many years, some sightings that give indirect evidence that some of them hunt bats at night. The evidence for some of them eating fish and birds is more direct." Perhaps the Van Meter cryptids visited the mine shaft to look for bats? According to a Net search, the most popular fish in the Raccoon River are Channel catfish, flathead catfish and walleye.

On November 25th Shane wrote to me as follows: I don't think that this is a bird, as close-up sightings never mentioned any feathers. If you shot a Sand Hill Crane or a Great Blue Heron at close range, with a shot gun, chances are it would be toast!

I really don't think this is an alien being that flew in from outer space, while not breathing oxygen and being bombarded with immense amounts of ultraviolet radiation and flying in zero gravity.

A Tulpa-thought-form likewise as with an alien, wouldn't need to descend a telephone pole like a parrot. They could just go back into thin-air. These kind of theories, don't really get us anywhere, as they are VERY difficult to prove.
Back here on Earth, the witnesses seem to be describing a pair- possibly a mated pair, of some type of flying reptile, similar to Pteranodons, that were using an abandoned coal mine as shelter. The horn on the head definitely signals a Pteranodon- a prehistoric animal, native to North America.

The light on the cryptid's horn could be the result of bioluminescent bacteria which are found in sea water, the surface of decomposing fish, in the guts of marine animals, and in freshwater and terrestrial animals. Sometimes, these bacteria form a symbiotic relationship with their host animal.

Thought forms also, don't leave physical evidence, such as the 3-toed-impressions in the soft ground, in Van Meter, that were plaster-casted. The plaster-casts were made of real tracks, not the ones someone carved out of a quarry. Those were completely different and obviously could not be plaster-casted!

Certain Pteranodon species were quite good at terrestrial movement. Pteranodons are considered quadrupedal, using both front and hind limbs for locomotion on the ground. Fossilized, quadrupedal, Pteranodon trackways confirm this.

The locomotion method described in the Van Meter reports is quite similar in this regard.

What would a Pteranodon be doing in Iowa, U. S. A., in the first place? It would be doing the same thing that thousands of other migratory birds do every spring and summer, migrating to the wetland habitats of the American Midwest, to take advantage of the population explosion of insects, plant life, amphibians, small mammals, and minnows that will sustain a nesting, piscivorous, flying reptile and it's brood.

Being primarily nocturnal, this cryptid would migrate under the cover of darkness, only to emerge in Midwestern wetlands at night, thus encountering the prolific, night-time insects, fish and amphibians. This nocturnal lifestyle would be a distinct advantage in keeping hidden from most human activity.

Modern reports of the Van Meter cryptid have continued until as recently as the summer of 2021.

Other Van Meter Cryptid facts include: a very tall animal with bat-like wings. Again, this matches some type of Pteranodon species quite well. This is my cryptid identikit for the Van Meter Cryptid, including a reasonable explanation of the reported phenomenon."

In conclusion, far from being paranormal entities, the Van Meter cryptids were, in our opinion, living pterosaurs, which stretch in their distribution from Papua New Guinea to the heart of the U.S.A. and elsewhere in the world. It is our hope that one day they will be removed from being seen as cryptids and placed within the realm of mainstream zoology.

THE 'X' FORMATION: AN ANALYSIS AND INTERPRETATION

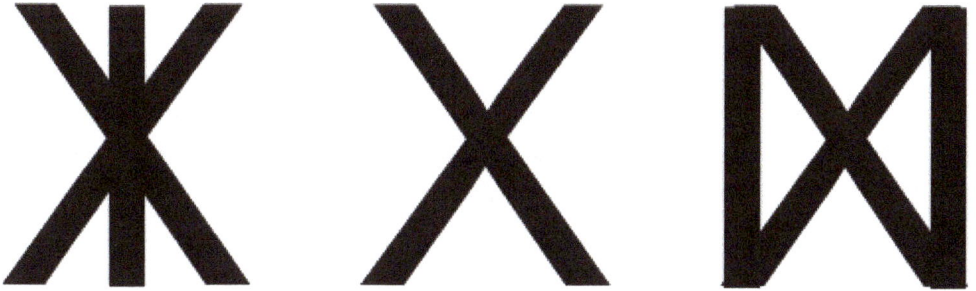

CONCEIVED, DEVELOPED, AND WRITTEN BY
SANJAY R. SINGHAL, R.A

TABLE OF CONTENTS

INTRODUCTION

Enthusiasts of Bigfoot (aka Sasquatch), among them experienced researchers and investigators, have documented numerous branch assemblies in areas of reported activity; sometimes labelled stick or branch structures, these constructions present indications of artifice, intelligent thought, and intention.

The interpretation and meaning of these assemblies is unclear or at the very least, unconfirmed. No one, to my knowledge, has ever witnessed their manufacture, although many have reported the discovery of same in certain and various sites, which appear and/or disappear on a regular basis, sometimes overnight or even within a few days.

Based upon the widespread proliferation and production of these assemblies, it is reasonable to infer in part that someone...or something...is putting these constructions together, for purposes we still do not know. While it is reasonable to suggest a human factor, there remain many outside such evidence. This document, accordingly, is an attempt to understand and to interpret one type of formation and to establish in part a means of identifying, interpreting, and comprehending other branch assembly types as well.

My source material has derived from my own investigations. It should be noted that, for the purposes of this document, my analyses and interpretations are entirely my own; they are not the work of others nor were they provided to me by others. Furthermore, it should also be noted that all graphics appearing in this document are my own work; I drew them myself and they belong to me.

I have stated that the analyses and interpretations presented herein are entirely the results of my own efforts, and so, I must also state that any mistakes, glaring or otherwise, are also my own. I'm hardly perfect, although I certainly try my best; the reader's gracious indulgence is humbly requested.

Respectfully submitted,
Sanjay R Singhal, RA

14 January 2022

BACKGROUND

I undertook my first 'official' paranormal/cryptid investigations at Salt Fork State Park and Woodbury Wildlife Area in southeast Ohio in May, 2007; both were memorable. At Woodbury, I experienced a terrifying, face-to-face encounter with a *very* large Bigfoot/Sasquatch, who positioned himself before my research partner and I, glaring furiously; it was and remains the most frightening experience of my life.

At Salt Fork however, despite some rather exciting activity, I observed and attempted to document a collection of strangely constructed, unusual branch assemblies observed in the thick woods and brush on a low hill near Hosak's Cave. It was my first investigation; I had never seen such fashions before and I daresay my architect's heart was roused by them. It was the beginning of a lifelong affair; it has continued to this day. The assemblies included a series of low, rounded, wigwam-like structures, a series of large, vertical constructions, and a long, horizontal branch, seemingly suspended above the ground. I also observed a feature of these assemblies for which I have attempted some initial interpretation, namely that *the size of a branch assembly may be directly related to the size of its maker.*

In the autumn of 2008, I observed my first upright 'X' formation in Emmet County,

Michigan; the impact of its beauty and the clarity of its construction have remained. Never before had I seen such a thing; even now, I recall asking myself: *who did this, and why*? It was observed in a dingle, positioned vertically between two upright, living trees; I recall other examples nearby, as well as several other branch assembly types. Upon my return to Emmett County a few weeks later, I was astonished to discover that many of these constructions had disappeared.

Since then, I have observed and documented various 'X' formations, ranging from Tennessee to Michigan, and from Illinois to California. All of them are amazing; all of them worthy of study and interpretation. Of necessity, I have included here only a few. But, as I have discovered within their analysis, there is a great deal more…than meets the eye.

DEFINITION

The 'X' formation, as I have defined it here, consists of two long, straight branches, usually stripped of their leaves and twigs and slanted vertically, one against the other, forming the letter 'X'. In some formations, portions of bark may have been removed from the branch component, revealing the brightly coloured sapwood; it is reasonable to suggest such manipulation(s) as deliberate.

In many formations, there is included an upright, living tree against which the formation rests and which serves as a support or fulcrum. This type of formation has been observed at Areas D and K in northern Michigan, in Orange County, California, in Lake County, Illinois, and also at Kelmarsh, in southern Wisconsin. In other formations, a pair of upright, living trees may be observed on either side: these may be considered stabiliser or brackets for the formation itself. This type has been observed in Cumberland County, Tennessee.

In only one location, Area D, have I observed an upright 'X' formation *freestanding*, ie not supported by other components but with its branches inserted forcefully into the ground. This assembly also featured a series of long, trailing vines interwoven about the interstices; it stood for many years.

A variant upon the 'X' formation is a composition of multiple, parallel components, usually comprised of a large, heavy branch and several smaller, thinner branches. This type has been observed at Area K in northern Michigan, and again at Kelmarsh.

In some (but not all) formations, smaller, curved branches, twigs, or vines may be interwoven about the junction of the two branches, as observed at Areas D and K, and also in Lake County, Illinois. Many formations also utilise a forked branch component and/or an arched branch component to maintain the other component in place; this has been observed in Cumberland County, Tennessee.

THE LETTER 'X'

The 'X' formation, despite its familiar appearance, should not be interpreted as anything akin to our own comprehension; it has many different meanings within our own history, society, and culture. While some of these may bear upon this document, it is not appropriate to infer any specific symbolism or consideration without further research and insight.

Historically, the letter 'X' is derived from the Greek letter *chi*, who borrowed it from the Phoenicians. This in turn made its way into the later Latin and Cyrillic alphabets, before finally becoming the 24th letter as we know it today in our own, modern English alphabet.

The Greek letter *chi*, written as X, is often used to represent the name of Christ (as in 'Xmas'); in combination with the letter *rho*, written as P, it is used to represent the risen Christ. Curiously, another combination of Greek letters, I and X, was used in the early Christian Church as another symbol of Christ, and is usually referred to as the IX or XI monogram. In contrast, a more secular view is provided in Plato's *Timaeus*, in which the *anima mundi* or the soul of the world, can be represented by two bands encircling the earth, which, at their intersection, form the letter X. These letters, alone or in combination, are compelling in their own right; while it may be reasonable to *suggest* a spiritual aspect to the 'X' formation, it is difficult to consider same without further evidence.

Prehistoric constructions, whether intended for religious and/or secular purposes, are usually found to be fashioned of stone or earth; they do not necessarily include other materials, wooden palisades, platforms, and/or fortifications notwithstanding. The famous Woodhenge, at Cahokia in southern Illinois, is an excellent example of such timber construction; continued research and investigation into similar sites may provide further insight. Vertical timber assemblies, akin to or resembling the 'X' formation, have not been documented at these sites.

In the mediaeval period, the 'X' formation appears in depictions of the 'wild man' or 'wood-wose'; it is presented as an attribute of same in a painting by Albrecht Dürer. This is a curious feature; it immediately suggests a correspondence with the 'wild man' of North America, and the association of same with intricately composed branch assemblies.

The 'X' formation and its construction, meaning, and purpose should not be assigned solely to Bigfoot/Sasquatch; there are other sources. The Boy Scouts of America learn and utilise a variety of trail markers, some of which incorporate long, straight branches or tree limbs; one of these is shaped like the letter 'X'. This is laid flat on the ground, and is a message to the scout 'Do not go this way'.

Another use of the letter 'X', albeit in the recent past and perhaps no longer as widespread, is the pictographic language of the hobo, or the tramp; there are many such examples. The letter 'X' appears within these symbols, but there appear to be conflicting meanings: I have found one source which interprets the letter 'X' as 'a good place' and another which interprets the letter 'X' as "man".

It is this last interpretation I have found to be of some interest; certainly the figure of a man, standing with his arms raised, bears a strong resemblance to the letter 'X'. An excellent example of this may be viewed in the 1968 film *Planet of the Apes*; there is a memorable scene wherein the crew of astronauts, led by Charlton Heston, discover a series of large scarecrows erected upon a high, rocky outcrop. The impression, immediate and pervasive, is of a series of large, upright 'X' formations!

One may suggest this as my 'Eureka!' moment; was it possible, I hypothesised, for the 'X' formations I had observed in the woods, to be a symbol for…Man?

THE 'X' FORMATIONS, IN SITU

EMMET COUNTY, MICHIGAN

I recall little of the first 'X' formation I observed in Emmet County, Michigan in 2008, other than its immense size, not only of its construction but of its members. It was clearly visible in the dingle, and I still remember, with a thrill, my first observation of it as I came over the hill. Although I did not at first consider the Bigfoot/Sasquatch responsible for its assembly, I nonetheless realised that I was looking upon an endeavour of some considerable artifice.

To the best of my recollection, the 'X' formation was quite tall, approximately twelve to fifteen feet high (12-15ft, or 3.66-4.57m), and of similar width; it was positioned (and possibly braced) between two upright, living trees. I do not now recall any component species; both appeared quite heavy and thick, approximately three to four inches diameter (3-4in, or 7.62-10.16cm).

Curiously, the formation did not appear to be prominently or visible sited; it had been erected in a small grove of trees at the bottom of the dingle, facing (approximately) south, and immediately adjacent to another curious feature: a large, moss-covered rock, whereupon I observed clumps of deer hair, with carefully cut, straight ends, and a scattered assortment of well-articulated bones, including a nearly intact spinal column.

AREA D

The first 'X' formation at Area D was discovered by my good friend and colleague Chuck Johnson* on a bitterly cold day in December 2011; it was erected at the edge of the clearing, adjacent to the meadow. It remained *in situ* for several years; sadly, I discovered it dismantled in March 2015. Nonetheless, upon each of my investigations at Area D, I documented the formation and its remarkable longevity; during the course of its life, it did not appear to change at all.

Upon my first examination, I observed a dense, twisted mass of wild grapevines wrapped about the formation; these were later removed. The formation itself was not large, approximately six to seven feet high (6-7ft, or 1.83-2.13m), but noticeably wider, eight to ten feet (8-10ft, or 2.44-3.05m).

The following features were observed:

♦ The formation was positioned at the transition from the clearing to the meadow, facing south across the meadow to the heavily wooded dunes beyond;
♦ The formation was clearly visible along the main trail;
♦ The formation comprised two (2) branches: one on the left, and one on the right;
♦ The left-hand branch presented a deep, brown-grey bark colour;
♦ The left-hand branch was heavy, long, and somewhat irregularly shaped;
♦ The right-hand branch was forked, with the left-hand branch inserted therein;
♦ The right-hand branch presented a pale, clear grey bark colour;
♦ The right-hand branch was somewhat more slender, with a slightly curved form;
♦ The right-hand branch was also forked, but this was not utilised.

These photographs present views and close-ups of the freestanding 'X' formation at Area D upon its initial discovery and investigation by Chuck Johnson and yours truly, 9 December 2011.

AREA K

One of the most intriguing 'X' formations observed at Area K was first discovered 5 January 2012; it had been erected in the woods, although I do not now recall the exact location. The formation was but one of an astonishing collection of branch assemblies and formations located throughout the property.

The formation itself was well-sized, approximately seven to nine feet tall (7-9ft, or 2.13-2.74m), but noticeably wider, approximately eight to ten feet (8-10ft, or 2.44-3.05m). Upon closer examination, I observed a gracefully arched, long red branch utilised at the junction of the two main components; it is reasonable to suggest this member provided some structural support.

The following features were observed:

♦ The formation was positioned in the front woods, with the farmhouse visible behind it;
♦ The formation appeared to face north (towards the farmhouse);

- ◆   The formation was clearly visible;
- ◆   The formation comprised two (2) main branches: one on the left, and one on the right;
- ◆   The left-hand branch presented a clear, grey bark colour;
- ◆   The left-hand branch was slender, long, and straight;
- ◆   The left-hand branch was forked, with the right-hand branch inserted therein;
- ◆   The right-hand branch presented a mottled, white bark colour with dark, bare patches;
- ◆   The right-hand branch was somewhat thicker and heavier, with a slightly curved form;
- ◆   The right-hand branch presented numerous branches remnant;
- ◆   The right-hand branch was inverted.

These photographs present views and close-ups of the 'X' formation at Area K, as initially observed 5 January 2012; in the first image, the farmhouse is visible through the bare trees

CUMBERLAND COUNTY, TENNESSEE

In the spring of 2013, while on a weekend field trip to Cumberland County, Tennessee, I observed an impressively large 'X' formation while riding; my mount would not stop, and I had to photograph the assembly in rather a hurry. No one else observed it; when we returned to the stables I shared my observation with the others. Later that afternoon, I returned and was able to examine it further.

The 'X' formation spanned a little-used hiking trail at a fork in same; the left-hand fork appeared to gently wind its way up into thickly wooded hills, while the right-hand fork proceeded directly up the hill slope. I estimated it to be approximately fifteen feet tall (15ft, or 4.572m). I took several photographs, with persons beneath, for scale purposes.

The following features were observed:

♦ The formation was positioned at the trail fork, (apparently) facing south towards the main trail as it ascended the hillside;
♦ The formation was visible for some distance along the main trail;
♦ The formation comprised two (2) branches: one on the left, and one on the right;
♦ The left-hand branch presented a pale red bark colour;
♦ The left-hand branch was heavy, long, and straight;
♦ The right-hand branch was inverted, and thrust deep into the ground;
♦ The right-hand branch presented a pale white bark colour;
♦ The right-hand branch was somewhat more slender, with a slightly curved form;
♦ The right-hand branch was forked, with the left-hand branch inserted therein.

These photographs present views and close-ups of the impressively large 'X' formation as first observed and investigated in Cumberland County, Tennessee, 13 April 2013. The first image was taken while riding; although I tried very hard, I could not stop my mount and snapped this photograph, almost turned sideways.

ORANGE COUNTY, CALIFORNIA

The 'X' formation in Orange County was first observed in late March, 2014, along a heavily-travelled street as I was driving with Mummie; I recall exclaiming rather loudly upon it! Mummie was tremendously interested; we returned a few days later and pulled off so I could examine it further. The shoulder was quite narrow; we could only stay a short time. Nonetheless, I was able to make some investigation of the formation, and to take several notes.

Unlike the other formations previously presented, the 'X' formation here was erected against an upright, living Wilson's Dogwood (*Cornus wilsoniana*), itself with a forked trunk, within which the two component branches were arranged. A slender, curved branch (species not identified) provided a fulcrum for the arrangement; the entire assembly was carefully and somewhat beautifully positioned. The formation itself was rather tall, approximately eight to nine feet high (8-9ft, or 2.44-2.74m), but somewhat narrower, approximately five to six feet (5-6ft, or 1.52 -1.83m).

The following features were observed:

♦ The formation was positioned at the edge of a steep, grassy bank, (apparently) facing south across a wide ravine to dense forest beyond;
♦ The formation was positioned above a stainless-steel, ribbed culvert, which emerged from beneath the road and opened towards the ravine;
♦ The formation was clearly visible along the highway;
♦ The formation comprised two (2) branches: one on the left, and one on the right;
♦ The left-hand branch presented a dull, red bark colour;
♦ The left-hand branch was long, straight, and somewhat slender;
♦ The left-hand branch was inverted;
♦ The right-hand branch presented a pale, clear grey bark colour;
♦ The right-hand branch was somewhat thicker and heavier, with a sawn-off end;
♦ The right-hand branch was inverted.

These photographs present views of the 'X' formation in Orange County as observed and investigated 30 March 2014; the first image was taken inside Mummie's car.

LAKE COUNTY, ILLINOIS

The 'X' formation at Illinois Beach State Park in Lake County, Illinois was first observed 15 November, 2015, although I did not investigate it until 25 November, 2015. I remain somewhat mystified that I did not observe the formation on my earlier visit to the park.

The formation, when I investigated it, was *enormous*; it was clearly visible across a wide expanse of marsh and its components, bleached white driftwood, contrasted vividly against the upright, living trees behind and beside it. It was very tall, approximately twelve to fifteen feet high (12-15ft, or 3.66-4.57m), and slightly wider, approximately fifteen to twenty feet (15-20ft, or 4.57-6.10m).

Similar to Orange County, the 'X' formation was erected against an upright, living Black Oak (*Quercus velutina*). A nearby sapling, possibly Black Oak as well, was curved into place, serving as a fulcrum for the arrangement; the entire assembly was carefully and really quite beautifully positioned.

The following features were observed:

♦ The formation was positioned at the edge of a thick belt of woods, (apparently) facing east across a wide expanse of marshland;
♦ The formation was clearly visible from the vehicle access road;
♦ The formation comprised two (2) branches: one on the left, and one on the right;
♦ Both branches were white, bleached driftwood;
♦ Both branches were long, and fairly straight;
♦ The right-hand branch was inverted, and thrust deeply into the ground;
♦ The left-hand branch featured remnant, forked branches at its upper end.

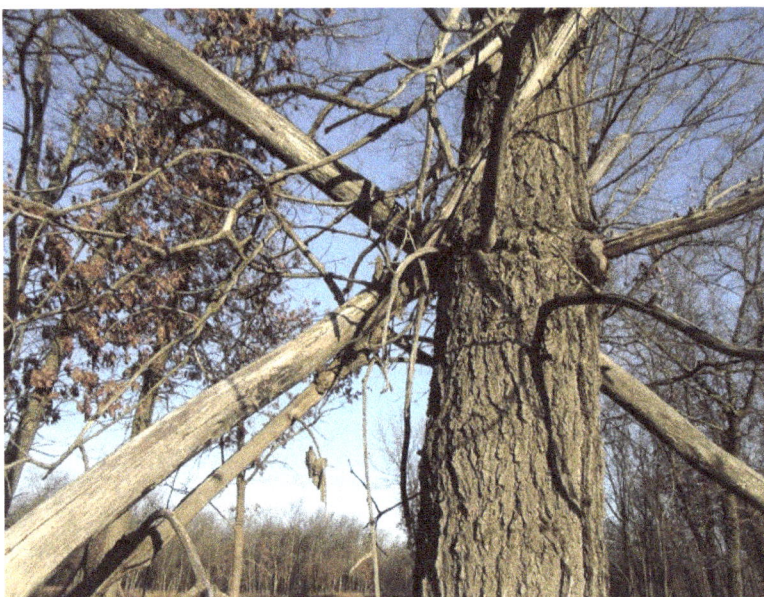

These photographs present views and close-ups of the very large 'X' formation at Illinois Beach State Park, as observed and investigated 25 November 2015. I still shudder when I think of crossing the swamp....

AREA D

The second 'X' formation at Area D was discovered by my good friend and colleague Chuck Johnson* in April 2016; however, neither he nor I realised, at the time, its true configuration. Curiously, the formation was incorporated into an older branch assembly, first observed in November 2011. Not until November 2016, when I returned to Area D alone, did I comprehend its full shape.

In similar fashion to the Orange County and Lake County formations, the component branches were positioned against an upright, living tree, in this location an American Beech (*Fagus grandifolia*) with numerous, small trunks. Numerous, trailing vines and long, slender branches were gathered about its base and components, as well. In remarkable contrast to other 'X' formations that I have investigated, all the components thereof appear to be of the same species, ie American Beech.

The formation itself was not large, approximately ten to twelve feet high (10-12ft, or 3.05-3.66m) and approximately the same width, as well.

The following features were observed:

♦ The formation was positioned at the edge of the heavily wooded dingle, (apparently) facing east across the clearing;
♦ The formation was only partially visible from the clearing;
♦ The formation utilised an older, remnant branch assembly;
♦ The formation comprised two (2) branches: one on the left, and one on the right;
♦ Both branches appeared to be American Beech (*Fagus grandifolia*);
♦ The left-hand branch was quite heavy and thick, and noticeably longer;
♦ The right-hand branch was forked, but did not appear to be utilised;
♦ The right-hand branch was noticeably more slender, with several remnant, smaller branches;
♦ The right-hand branch was inverted.

These photographs present views and close-ups of the 'X' formation at Area D as observed and investigated on 5 November 2016; it was only upon this enquiry that I realised its overall shape and configuration.

SUMMARY

Some initial comparisons may be derived from the examples of 'X' formations presented in this document; however, no definitive statements should be considered. Further research and investigation of a broader range of formations, in a variety of locations and settings, is required in order to provide a fuller and more useful analysis and comprehension of this type of assembly. Some examples, of other 'X' formations researched and photographed by Mr Rick Reles of Wisconsin, are included in the Appendix.

Nonetheless, the following features may be considered practicable for future investigation of EC-related branch assemblies and configurations:

The 'X' formation is usually positioned at the edge of the woods, and appears to be facing towards a clearing, a road, or other areas of human passage and/or activity;
The 'X' formation is usually comprised of two (2) main components, but may incorporate a lesser, third component, ie a gently curved branch or sapling;
The 'X' formation appears to remain *in situ* for an extended period;
The main components are usually contrasting; ie different bark colours, and sometimes with noticeably different textures;
One or more of the main components may be inverted;
One or more of the main components may be forked, and utilised structurally;
One of the components may be heavier and thicker, than the other;
Inversely, one of the components may be longer, and more slender, than the other;
One of the components may be somewhat more rugged and non-linear in form.

CONCLUSION

The placement of 'X' formations, if EC-related, may be considered a marker for human passage and/or activity, and as such may also serve as a boundary adjacent to same. Although a human initiative may not be fully eliminated, the possibility of such labours must be considered improbable at best, based not only upon the size of the 'X' formations as observed, but also of the effort necessary to assemble them *in situ*, with component branches not necessarily original to their location.

Further aspects of the 'X' formations reveal a degree of constructive skills not immediately related to human activity, in particular the use of inverted branches driven forcefully into the ground and/or the use of heavier, thicker branch components to stabilise and maintain the assembly *in toto*. The utilisation of a slender, curved branch to provide additional structural support indicates a comprehension of the tensile properties of wood.

Throughout this document, I have presented my descriptions of the 'X' formations *as I observed them*, ie the left-hand branch, and the right-hand branch, and so on. It is necessary however, to perceive these structures not with human eyes, but with those of their engineers: in so doing, we must observe them…from the shadows.

Seen from the woods, peering out through the branches, the 'X' formation is observed in silhouette; it becomes in effect a giant shape upon the landscape, a clear and visible marker of the passage of others. One may suggest that the form of this assembly, the letter 'X', is intended as a symbol for us.

Here, there be *humans*.

APPENDIX

These photographs were taken by Mr Rick Reles of Wisconsin, and are included here with his kind permission; they present, in order, a series of 'X' formations in the Kettle Moraine State Forest, Northern Unit, Newport State Park, Myakka River State Park, and the Kettle Moraine State forest, Southern Unit.

NOTES

1. My former colleague, Joshua Paul 'JP' Smith, once told me of some video footage which presented evidence of Bigfoot/Sasquatch in the process of fashioning a branch assembly; however, he did not share this with me, and I do not have any further information regarding same. Had he done so, it no doubt would have aided my efforts immensely.
2. Joshua Paul 'JP' Smith died 18 July, 2015.
3. One of my former research sites, Area K in northern Michigan, was famous for the rapid construction of such assemblies; in many instances, assemblies documented one day would be gone the next, or new assemblies would be observed, which had not been there the day before. Regrettably, despite such activity, no focussed research and/or documentation of same was ever undertaken.
4. The destruction of nest-type assemblies, almost overnight or at least immediately upon their discovery, has been documented at three (3) of my own research sites: Areas D & K in northern Michigan, and Area R in Illinois. Although some inference of regional and/or local practise may be made, further investigation is required.
5. Sanjay R Singhal. *Field Report 05.18.2007: Salt Fork State Park.* Beyond The Forest. https://beyondtheforestblog.wordpress.com/2014/08/03/field-report-05-18-2007-salt-fork-state-park/. 3 August 2014. Web. Accessed 25 February 2017.
6. Ibid. *Field Report 05.19.2007: Woodbury Wildlife Area.* Beyond The Forest. https://beyondtheforestblog.wordpress.com/2014/08/10/field-report-05-19-2007-woodbury-wildlife-area-2/. 10 August 2014. Web. Accessed 25 February 2017.
7. Similar branch constructions have been observed at Woodbury Wildlife Area, as well as Area K.
8. Regrettably, I did not document these assemblies at all thoroughly or well. Sigh....
9. I have noted certain and various groupings of adult/child branch assemblies in other areas of Salt Fork, as well as Area D in northern Michigan. It is reasonable to suggest, in context, other similar groupings at Area K, but this relationship was not recognised at the time of my investigation(s) there.
10. This topic would make for an excellent analysis as well.
11. Ibid, Singhal. *Field Report 10.25.2008: Emmet County.* Beyond The Forest. https://beyondtheforestblog.wordpress.com/2014/09/02/field-report-10-25-2008-emmet-county/. 2 September 2014. Web. Accessed 25 February 2017.
12. See previous footnote regarding assembly destruction.
13. I do not, regrettably, have any photographs of these assemblies; I have no idea what happened to them.
14. This is a commonly observed feature of branch assemblies, regardless of type.
15. Richard L Venezky. *The Structure of English Orthography.* The Hague: Mouton & Company; 1970.
16. The slanted-x formation is a variant of this type.
17. Research sites in northern Michigan are letter-coded: B, D, F, H, K, and P.
18. Research sites in southern Wisconsin are name-coded: Kelmarsh, Hammerfield, Penrose, Rutland Gate, et al.
19. BF Cook: *Greek Inscriptions: Reading the Past.* Berkeley and Los Angeles: University of California Press; 1998.

20. Andrew Cross. *The Development of the Greek Alphabet within the Chronology of the ANE.* Calgary, Alberta: University of Calgary; 2009.
21. Stephen R Fischer. *A History of Writing.* London: Reaktion Books; 2001.
22. Florian Coulmas. *The Blackwell Encyclopaedia of Writing Systems.* Oxford: Blackwell Publishers; 1999.
23. RC Sproul. *What Does the X in Xmas Mean?* Ligonier Ministries. http://www.ligonier.org/blog/why-is-x-used-when-it-replaces-christ-in-christmas/. 19 DBigfoot/Sasquatchember 2016. Web. Accessed 14 March 2017.
24. Alva William Steffler. *Symbols of the Christian Faith.* Grand Rapids, Michigan: WB Eerdmans Publishing; 2002.
25. Frederick R Webber. *Church Symbolism (2nd Edition).* Toronto: Omnigraphics; 1980.
26. Plato. *Timaeus.* 360BC.
27. Wisconsin DNR Author(s). *Aztalan State Park: History.* Wisconsin Department of Natural Resources. http://dnr.wi.gov/topic/parks/name/aztalan/history.html. 2015. Web. Accessed 20 August 2015.
28. Ohio History ConnBigfoot/Sasquatchtion Author(s). *Serpent Mound.* https://www.ohiohistory.org/visit/museum-and-site-locator/serpent-mound. 2017. Web. Accessed 15 March 2017.
29. Native Stones Author(s). *Cairns.* Native Stones. http://www.nativestones.com/cairns.htm. 2005. Web. Accessed 16 March 2017.
30. Dean Snow. *Archaeology of Native North Americas.* Upper Saddle River, New Jersey: Prentice Hall; 2010.
31. Sally AK Chappell. *Cahokia: Mirror of the Cosmos.* Chicago: University of Chicago Press; 2002.
32. Increase A Lapham. *Antiquities of Wisconsin.* Washington, DC: Smithsonian Institution; 1855.
33. Robert Withington. *English Pageantry: An Historical Outline.* Cambridge: Harvard University Press; 1918.
34. David J Daegling. *Bigfoot Exposed: An Anthropologist Examines America's Enduring Legend.* Lanham, Maryland: Altamira Press; 2004. There are numerous accounts of the 'wild man' or 'hairy man' among the indigenous peoples of North America, with varying attributes, personalities, and mythologies.
35. Emil Krén and Daniel Marx. *Portrait of Oswolt Krel.* Web Gallery of Art. http://www.wga.hu/html_m/d/durer/1/02/12krel.html. Date Unknown. Web. Accessed 16 March 2017.
36. Sometimes referred to as 'Bigfoot' or 'Sasquatch', although neither of these terms is favoured.
37. The Inquiry Author(s). *Trail Signs: Traditional.* The Inquiry. http://www.inquiry.net/outdoor/skills/b-p/signs.htm. 15 October 2016. Web. Accessed 16 March 2017.
38. The Free Dictionary Author(s). *Hobo Convention.* The Free Dictionary. http://forum.thefreedictionary.com/postst117721_Hobo-Convention.aspx. 8 August 2015. Web. Accessed 16 March 2017.
39. Ernest Thompson Seton. *Tramp Signs.* The Inquiry. http://www.inquiry.net/outdoor/skills/seton/blazes.htm. 15 October 2016. Web. Accessed 16 March 2017.
40. Franklin J Schaffner. *Planet of the Apes.* Los Angeles: Twentieth Century Fox; 1968.
41. Ibid, Singhal. *The Chisholm Assembly: Analysis and Followup.* Beyond The Forest. https://beyondtheforestblog.wordpress.com/2015/08/01/the-chisholm-assembly-analysis-and-followup/. 1 August 2015. Web. Accessed 17 March 2017. I have long

theorised that branch assemblies, if considered to be erected by the Bigfoot/Sasquatch, may not necessarily be intended for us…but may be…*about* us.

42. Ibid, Singhal. *Field Report 10.25.2008: Emmet County.*
43. It was later suggested by the landowner that poachers had been hunting deer on the property, and skinning them in the forest, leaving the bones and other refuse in this area.
44. Ibid, Singhal. *Field Report 12.09.2011: Area D.* Beyond The Forest. https:// beyondtheforestblog.wordpress.com/2014/12/27/field-report-12-09-2011-area-d/. 27 DBigfoot/Sasquatchember 2014. Web. Accessed 18 March 2017.
45. Ibid. *Field Report 03.21.2015a: Area D.* Beyond The Forest. https:// beyondtheforestblog.wordpress.com/2015/10/11/field-report-03-21-2015a-area-d/. 11 October 2015. Web. Accessed 18 March 2017. Curiously, the components of the 'X' formation were still in the clearing, albeit scattered about.
46. Ibid. *Field Report 08.31.2012: Area D.* Beyond The Forest. https:// beyondtheforestblog.wordpress.com/2015/05/23/field-report-08-31-2012-area-d/. 23 May 2015. Web. Accessed 18 March 2017. Most of the wild grapevines had been removed by the end of August 2012.
47. Ibid.
48. I remain somewhat embarrassed that Chuck Johnson observed this formation before I did!
49. It is my own theory that Bigfoot/Sasquatch-related branch assembly components include contrasting bark colours, thicknesses, textures, and species; I have documented this feature in numerous locations.
50. It is my own theory that the 'topmost' branch assembly component is larger and heavier; the weight of this branch stabilises and thus maintains the formation *in situ*.
51. Carl H Tubbs and David R Houston. *American Beech (Fagus grandifolia).* USDA Forest Service Sylvics Manual Volume Two. http://www.na.fs.fed.us/pubs/ silvics_manual/volume_2/fagus/grandifolia.htm. Date Unknown. Web. Accessed 3 August 2014.
52. Ibid, Singhal. *Field Report 01.05.2012: Area K.* Beyond The Forest. https:// beyondtheforestblog.wordpress.com/2015/01/04/field-report-01-05-2012-area-k/. 4 January 2015. Web. Accessed 18 March 2017. Per my notes from this investigation, the formation was located in the 'front' woods.
53. Several heavy, thick belts of forest surrounded the farmhouse at Area K; most of the branch assemblies were constructed in the 'front' woods, immediately south of same, although others were also observed in the 'north' and 'east' woods.
54. Casey J Mansfield. *Does the Degree of Curvature Affect the Strength of Wooden Arches?* California State Science Fair 2004 Project Summary. https://www.usc.edu/CSSF/ History/2004/Projects/J1811.pdf. February 2004. Web. Accessed 22 January 2015. The tensile property and strength of a curved piece of wood is astonishing, and can be quantifiably measured.
55. For the purposes of this document, it should be noted that the formation was observed in winter, when the trees were bare; it is reasonable to suggest that, in the summer, it may have been less visible. Regrettably, I do not recall investigating this formation during the summer of 2012, before the farm was sold.
56. It is my own theory that Bigfoot/Sasquatch-related branch assembly components include contrasting bark colours, thicknesses, textures, and species; I have documented this feature in numerous locations.
57. Ibid, Tubbs and Houston. *American Beech (Fagus grandifolia).*

58. It is my own theory that the 'topmost' branch assembly component is larger and heavier; the weight of this branch stabilises and thus maintains the formation *in situ*.

59. Similar branches, inserted vertically into the ground, have been observed in Comins, Michigan (Oscoda County); it is reasonable to suggest, in context, a regional variant of this behaviour, although further research and investigation is required.

60. Amy Bennett. *New York Bigfoot Society Field Report: November 2013*. New York Bigfoot Society. http://www.newyorkbigfootsociety.com/fieldreportnov2013. 2014. Web. Accessed 18 January 2015. Reports of saplings inserted forcefully into the ground are somewhat common to Bigfoot/Sasquatch-related branch assemblies.

61. I continue to be amazed that no one else observed the 'X' formation while we were riding.

62. It is my own theory that Bigfoot/Sasquatch-related branch assembly components include contrasting bark colours, thicknesses, textures, and species; I have documented this feature in numerous locations.

63. It is my own theory that the 'topmost' branch assembly component is larger and heavier; the weight of this branch stabilises and thus maintains the formation *in situ*.

64. Similar branches, inserted vertically into the ground, have been observed in Comins, Michigan (Oscoda County); it is reasonable to suggest, in context, a regional variant of this behaviour, although further research and investigation is required.

65. Ibid, Bennett. *New York Bigfoot Society Field Report: November 2013*.

66. Ibid, Singhal. *Field Report 03.30.2014: Orange County*. Beyond The Forest. https://beyondtheforestblog.wordpress.com/2015/09/05/field-report-03-30-2014-orange-county/. 5 September 2015. Web. Accessed 20 March 2017.

67. As of December 2016, the formation remains *in situ*.

68. I am extremely blessed to have a Mummie so interested!

69. Sheffield's Seed Company Author(s). *Wilson's Dogwood (Cornus wilsoniana)*. Sheffield's Seed Company. https://sheffields.com/seeds/Cornus/wilsoniana. 2013. Web. Accessed 4 September 2015.

70. Ibid, Mansfield. *Does the Degree of Curvature Affect the Strength of Wooden Arches?*

71. I remain somewhat embarrassed that Chuck Johnson observed this formation before I did!

72. It is my own theory that Bigfoot/Sasquatch-related branch assembly components include contrasting bark colours, thicknesses, textures, and species; I have documented this feature in numerous locations.

73. Don Minore. *Western Red Cedar (Thuja plicata)*. USDA Forestry Service Sylvics Manual Volume One. http://www.na.fs.fed.us/pubs/silvics_manual/Volume_1/thuja/plicata.htm. Date Unknown. Web. Accessed 4 September 2015.

74. It is my own theory that in Bigfoot/Sasquatch-related constructions the 'topmost' branch assembly component is larger and heavier; the weight of this branch stabilises and thus maintains the formation *in situ*.

75. Ibid, Singhal. *Field Report 11.15.2015: Illinois Beach State Park*. Beyond The Forest. https://beyondtheforestblog.wordpress.com/2015/12/13/field-report-11-15-2015-illinois-beach-state-park/. 13 December 2015. Web. Accessed 20 March 2017.

76. Ibid. *Field Report 11.25.2015: Illinois Beach State Park*. Beyond The Forest. https://beyondtheforestblog.wordpress.com/2015/12/20/field-report-11-25-2015-illinois-beach-state-park/. 20 December 2015. Web. Accessed 20 March 2017.

77. As of October 2016, the formation remains *in situ*.

78. Ibid. *Field Report 11.04.2015: Illinois Beach State Park.* Beyond The Forest. https://beyondtheforestblog.wordpress.com/2015/12/06/field-report-11-04-2015-illinois-beach-state-park/. 6 December 2015. Web. Accessed 20 March 2017.

79. An appropriate inference may be made....

80. This is a critical feature; see below.

81. Ivan L Sander. *Black Oak (Quercus velutina).* USDA Forest Service. http://www.na.fs.fed.us/pubs/silvics_manual/volume_2/quercus/velutina.htm. Date Unknown. Web. Accessed 11 May 2015.

82. In contrast to the Orange County formation, however, the assembly here had been erected *behind* the upright, living tree; this perspective is based upon an east-facing view, of course.

83. Ibid, Mansfield. *Does the Degree of Curvature Affect the Strength of Wooden Arches?*

84. How do I know it was a wide expanse? I walked across it...and thoroughly soaked my feet, in the process!

85. It is my own theory that Bigfoot/Sasquatch-related branch assembly components include contrasting bark colours, thicknesses, textures, and species; I have documented this feature in numerous locations.

86. Similar branches, inserted vertically into the ground, have been observed in Comins, Michigan (Oscoda County); it is reasonable to suggest, in context, a regional variant of this behaviour, although further research and investigation is required.

87. Ibid, Bennett. *New York Bigfoot Society Field Report: November 2013.*

88. Ibid, Singhal. *Field Report 04.07.2016a: Area D (Parts 1 & 2).* Beyond The Forest. https://beyondtheforestblog.wordpress.com/2016/04/20/field-report-04-07-2016a-area-d-parts-1-2/. 20 April 2016. Web. Accessed 21 March 2017.

89. Ibid. *Field Report 11.13.2011: Area D.* Beyond The Forest. https://beyondtheforestblog.wordpress.com/2014/12/26/field-report-11-13-2011-area-d/. 26 December 2014. Web. Accessed 21 March 2017.

90. Ibid. *Field Report 11.05.2016: Area D.* Beyond The Forest. https://beyondtheforestblog.wordpress.com/2017/01/10/field-report-11-05-2016a-area-d/. 10 January 2017. Web. Accessed 21 March 2017.

91. Ibid, Tubbs and Houston. *American Beech (Fagus grandifolia).*

92. It is reasonable to suggest that, in context, this profusion of upright, vertical trunks may have served as an effective camouflage, disguising the overall outline of the formation upon our first investigation.

93. Ibid, Singhal. *Field Report 11.13.2011: Area D.*

94. It is my own theory that in Bigfoot/Sasquatch-related constructions the 'topmost' branch assembly component is larger and heavier; the weight of this branch stabilises and thus maintains the formation *in situ.*

95. The *first* formation observed at Area D and the formation observed at Area K, however, may be considered exceptions.

96. The formations observed in Orange County and Lake County are notable examples.

97. The formations observed in Orange County, Lake County, and at Area D are notable examples.

98. The formations observed at Area D, Area K, Cumberland County, and Orange County are notable examples.

99. The formations observed at Area K, Cumberland County, Orange County, Lake County, and Area D (2nd formation) are notable examples of this feature.

100. The formations observed at Area D, Area K, and Cumberland County are notable examples.

101. The formations observed at Area D (1st & 2nd formations), Area K, Cumberland County, and Orange County are notable examples of this feature.
102. The formation at Lake County, utilising large pieces of driftwood, is a notable example.
103. Wisconsin DNR Author(s). *Kettle Moraine State Forest Northern Unit.* Wisconsin Department of Natural Resources. http://dnr.wi.gov/topic/parks/name/kmn/. 2015. Web. Accessed 12 September 2015.
104. Ibid. *Newport State Park.* Wisconsin Department of Natural Resources. http://dnr.wi.gov/topic/parks/name/newport/. 2017. Web. Accessed 21 March 2017.
105. Florida DEP Author(s). *Myakka River State Park.* Florida Department of Environmental Protection. https://www.floridastateparks.org/park/Myakka-River. 2017. Web. Accessed 21 March 2017.
106. Wisconsin DNR Author(s). *Kettle Moraine State Forest Southern Unit.* Wisconsin Department of Natural Resources. http://dnr.wi.gov/topic/parks/name/kms/. 2015. Web. Accessed 14 January 2016.

X MARKS THE SPOT
Alex Mistretta

Jaqueline Roumeguere Eberhardt was a well known and respected anthropologist in a world far from Maasai villages and the deep forests of Kenya and Tanzania. Here however, among the Maasai people she was simply 'Jaqueline', a friend and an accepted member of the community. Before all that she had obtained a Masters degree at the University of Johannesburg, moved to France and obtained two Doctorates and then went on to head the prestigious CNRS, the French National Centre for Research. She was firmly entrenched in academia, but the call of Kenya was too difficult to ignore, and these strange creatures she called X changed everything.

She would step away from the limelight, partly due to the hostility towards the possibility of X. X was the term she used for all Yeti/Bigfoot types in Africa, but also for more human Hominins that appear to live in the forests and jungles of Kenya and Tanzania.

She labelled them:

X1 : This is the most common type, similar to Bigfoot, Yetis and Russian Almastys but with apparently more variation in size, especially towards the smaller scale... The face is human looking, somewhat devoid of hair, although with thicker features. X 1 sometimes uses a club or a branch while hunting, especially for buffalos. While they rarely attack domestic animals, wild buffalos appear to be one of their primary source of meat. Also, they routinely drink the blood of slain buffalos.

They are non aggressive towards humans, but have a certain curiosity towards children. They apparently at times grab children, and sometimes adults, to observe them for a little while and then let them go. They do vocalize among themselves, but whether that constitute a bona fide language is unknown. JRE clearly felt that they were hominins, perhaps a continuation of Australopithecines or Homo erectus types. Isabell, her daughter, confirmed this hypothesis to me. X1 lives in the forest.

X2: X2 is human and not covered in hair, and with a light colored skin. X2 is larger on average than local Africans. Considerably so in fact, and very aggressive and considered very dangerous. They have a strange habit of getting excited by fire and rushing fires to put them out scatter the ashes all over. Very little else is known about them, but they are thought to live in caves.

X3: X3 is an isolated case of what appears to be a member of the Maasai tribe that lived alone in the forest. He was old, so since these reports are from the late 1970's, presumably now dead.

X4: Also appears to be a case of a tribe of humans that live further out in the forest. X4 is small, with a clear language and quite possible an unknown Pygmy tribe. Not many sighting of this type of X and like X3, I think these may be modern humans. Since X 4 eats raw meat and they are small in stature, JRE speculated that they may

be a relative of early humans, but when reading her book, I get the impression that she was much less sure of having a hypothesis for X4 and X3, especially as compared to X1 and X2.

X5: Clearly human, and like X2, X3 and X4 uses artifacts, but is known from only one case, but an important one. X5 is the one that left the famous bow and arrow behind, which I'll discuss a little later in more detail. His relationship to X2 or X3 or X4 is unknown, and because of that and the uniqueness of the bow and arrow, JRE labelled this one case separately as X5.

What's important is that she didn't obtain her data from strangers where cultural translation could interfere with the interpretation of that data. She had been living among the people since 1966 and as such she obtained her information and stories from friends, people she knew, people she learned to share a culture with. That makes all the importance in the world. This was the best kind of Anthropology, one trained in collecting evidence and interpretation but also first hand insider knowledge.

As such she had a unique advantage in being able to receive stories that normally would remain internal and It was on such an occurrence that a 30 year old Maasai she had known for quite sometime noticed a magazine she was reading, *Time* magazine to be precise, from 1977, with a representation of *Homo habilis* on the cover next to Richard Leakey. Leakey ironically would later become very critical of JRE's hypothesis that an archaic species of man had survived to the modern day in African forests. Here, her friend, a former Maasai warrior taking interest in the depiction of *Homo habilis* looks at JRE and says "you know I've seen him, but he is much more powerful,… and my friend had even been held captive by one".

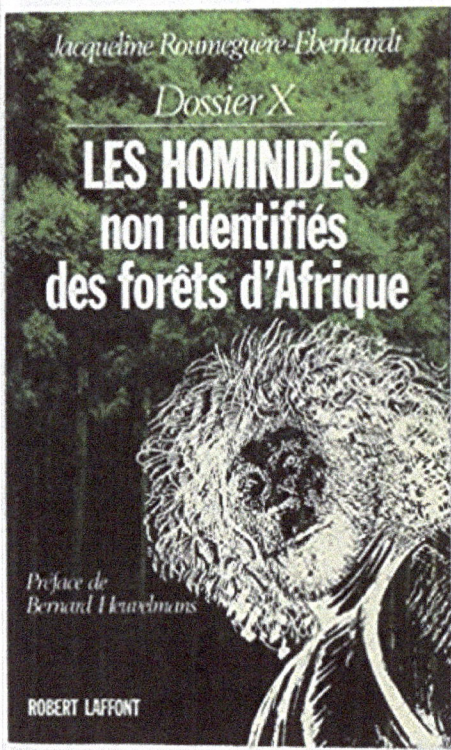

THE SPECIES

Intrigued, she started enquiring and collecting more and more stories of up close encounter with these strange dwellers in the forest and quickly realized that she might be dealing with several different types, as many as five type of X as we have already discussed. The culmination of her research would result in her book *Les Hominides Non Identifies des Forets D'Afrique* which was only published in France.

In the book she tells of close to 80 encounters with - as we have seen - both Bigfoot types and human types that can only be described as similar to ancestral humans. The book doesn't tell all though. I contacted her daughter, Isabelle, who informed me she had collected 133 accounts. Up close sightings of X1 included the sighting of a female picking bugs off the male and eating them in typical ape behavior, but the difference is that they walked away on two feet. There were numerous close encounters from hunters apparently out for the same prey as X, and even a dead specimen another hunter found lying in the forest. The descriptions were pretty typical, covered with hair, naked, no language, very robust and bipedal. They were also other tales of what appears to be more human versions so X2, but still not belonging to any known tribe. The later, Isabelle informed me, were certainly human, but didn't appear to have a language and use primitive tools, and perhaps most peculiar of all, they have a habit of covering themselves in ash. I didn't have the chance to further inquire about the 'covered in ash people', but I assume this refers to x2.

A quick note here in terms of the usage of the term 'primitive', which unfortunately resulted in a rift between myself and Isabelle. 'Primitive' as I learned to use the term as an anthropology student has no negative connotations. It does not infer inferior. It is simply a term designed to designate a tool culture or specific physical attributes from the distant past that no longer occur in modern *Homo sapiens.*

As such, descriptions of X1, which to JRE resembled *H habilis* or *H erectus*, or even the Australopithecines as Isabelle hypothesized, I would describe as more primitive to the more human X2. The later versions of X, whether X2, X3, X4, X5 whose cultures uses elephant sinew bow and arrow (X5), cover themselves in ash, use knives and clubs to kill buffalos in order to drink their blood and raw internal organs, would then be called more primitive from a cultural standpoint as compared to other tribes who use antelope sinew, metal for their arrows and cook their food.

WHAT IS X

X has become a somewhat ubiquitous term for mysterious Hominids and Hominins in the region from the ash covered people to the more Bigfoot types to the more gracile sightings resembling early primitive species such as australopithecines and more recent human relatives. You cannot separate the story of X from the story of human evolution. I repeat you cannot separate the story of X and the story of human evolution. The implications are unavoidable. The similarities with hominins that once co-existed together in this place before branching out into a bigger world is unavoidable. Isabelle Roumeguere herself told me that the comparison with the Australopithecines are unavoidable. Yes it makes me uncomfortable on some level. We just don't have enough data. Hopefully my upcoming expedition, provided it actually happens this time, can to a degree change all that.

The easier assumption based on what we know of Bigfoot and also the Yeti has always been labelling them apes. And certainly based on observed behavior such as bluff charges, ground nests and physical attributes such as the sagittal crest, you can make a case for that. All three attributed to the North American Bigfoot makes a case for ape ancestry. Furthermore the easier assumption for Orang Pendek, the diminutive "Bigfoot" of Sumatra for example is ape since they do not appear to have language or a tool culture. And certainly there are similarities between some of the sightings in the Congo, Cameroon and Tanzania of the Agogwe and Kakundakari and Sumatra's Orang Pendek.

On the other hand, certain traits attributed to Bigfoot such as its vocalization and much of the history of the Russian, Mongolian and Central Asian Yeti suggests a more human, Hominin, creature.

But X is different, and that's why it is so important. Differences and uniqueness are sometime just as important clues as commonalities. Granted, part of that is that X doesn't refer to just one thing and since aside from JRE's book, we have a limited amount of data, and certainly the data between X1 and X2 may at times overlap and be amalgamated. Off topic, I do think this is also an issue with Africa's other big mystery, Mokele-mbembe and Emela-ntouka (Chipekwe). This is something we often face as Cryptozoologists, but it shouldn't deter us from forming reasonable hypotheses and taking a stand.

SOME SIGHTINGS OF NOTE

Here I want to cover some of the more spectacular sightings of X.

1- One eyewitness who was hiding behind a tree saw two X, one male and one female. He said despite the fact they were hairy all over, they looked like people.

The female was removing bugs from the male's fur and either eating or crushing them. During this time, there was some type of vocalization between the two. In any case at some point they stood up and walked away bipedally. As the eyewitness quietly left the area, fearing for their lives dues to the size and robust aspect of X, although the female was slightly more slender, they had to go by a stream and there they saw five juveniles, much smaller than the other two but with lighter colored hair; light brown to red.

2-   Another eyewitness, who had been captured by X, described it as being covered in hair except for in the facial area, where it had manlike facial features, but with low forehead kind of like a baboon.

3-   A Swedish CEO was walking down a path in the forest when he encountered X. Stunned, as he had never heard of such a thing, he ceded the passage to this being who looked at him as they both continued going their separate way.

4–   More recently, there had been sightings along the Dja river in Cameroon. This is Mokele-mbembe territory. Mokele-mbembe is my big love, having tried to launch an expedition in the past, I've been obsessed with resurrecting that and as such have been paying attention to what's going on there. In any case, I accidently found an interview done in France about a recent expedition along the Dja and they mentioned sightings there of a Hominid similar to X1 and the Kakundakari.

5-   One eyewitness came upon a dead female X. She was covered in hair with elongated breasts. The breasts were hairless as was much of the face. The facial features appeared human like but with noticeably large teeth.

BOWS AND ARROWS

My interest and my focus is obviously X1 and X2 and the larger picture they represent. Both have enough traits reminiscent of earlier Hominins that we're talking about a subject that can revolutionize human history. Also the link to the Out of Africa migration and thus to a likely explanation for the Almas and Almasty of Russian, Mongolia and Central Asia is inescapable and although a little beyond the scope of this article, it does present a fascinating bigger picture.

Also, is there a link with the Yeti of the Pyrenees? I was initially skeptical of sightings in the Pyrenees, but recent research and events have changed my mind. I'm hoping to visit the Pyrenees pretty soon, possibly as early as the Summer or Fall 2022. The bottom line is there's a bigger picture at work here.

I also want to focus on the bow and arrow, although I'm pretty sure and JRE confirms this, that they belong to a human. How modern of a human, remains to be seen, but I would be surprised if X5 turns out to be something more. That said, X5 isn't the only X with artifacts, so having something tangible certainly helps, especially when attempting to secure funding for an expedition. Also, we know the bow and arrow exists, I know where they are.

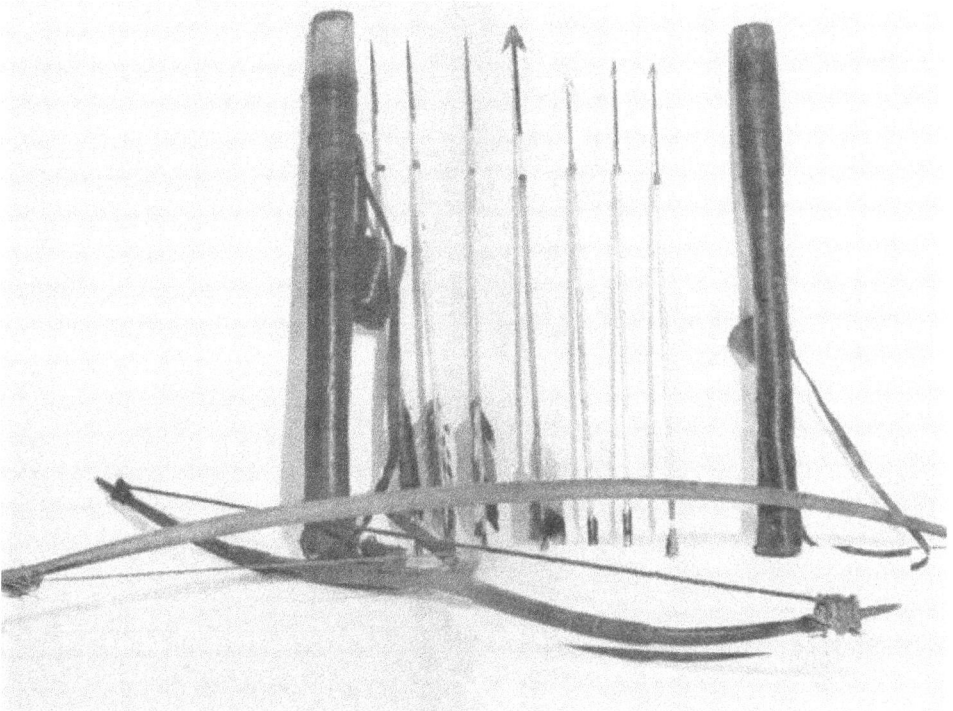

Still the bow and arrows displays a more primitive technology as compared to bows and arrows found as compared to any known tribes in the region and JRE could not match the level of technology, type, shape with any known tribe. In colloquial terms, and a somewhat dated anthropology modality, the bow and arrows were stone age technology, and it didn't belong in modern day Africa. JRE felt that they, in all likelihood, belonged to an unknown primitive tribe of *Homo sapiens*. Whether there is a link with the more primitive X remains to be seen and this is why further study is necessary in the forests of Kenya.

A COMPLICATION OF PERSPECTIVE

Things get further complicated once we take an overall look at the mysterious "Bigfoot" like situation in the region and Africa in general. JRE as a professional Cultural Anthropologist, conducting years of research and interviewed countless eyewitnesses and shared her data in her book *Les Hominide Non Identifies des Forets D'Afrique*. But here's the thing that's really important; she was a Cultural Anthropologist first and not a Physical Anthropologist, and the difference resides in how you share your data. This is important.

My Bachelor of Science degree, for example, is in Physical Anthropology so I look at everything from an evolutionary standpoint and look for the genetic relationship between species. So when I deal with Bigfoot or the Yeti or in this case X, I infer from the evidence and data their place in taxonomy. I infer speciation. Are they Hominids, meaning apes or humans or ancestors/relatives of apes or humans? Or, are they Hominins, humans and direct relatives and ancestors? What can I infer from their behavior or from the evidence that we have?

Jaqueline Roumeguere-Eberhardt is obviously trained in that as well, and obviously a lot more capable than I am as an Anthropologist, but let's remember that she didn't have access to all the latest genetic discovery and the revolution that has taken place in human evolutionary biology in recent years.

She didn't know about *Homo floresiensis*, Denisovans, *Homo naledi* and the late survival of *Homo erectus*. We now know beyond a shadow of a doubt, that modern humans co-existed with multiple Hominin species until quite recently, many of which had a combination of modern traits and very primitive traits. Meaning, that numerous species similar to X not only now exist in the recent fossil record, but co-existed with modern humans and shared genes that we can now trace back. This changes the game drastically.

I've been thinking about an expedition for decades, and I finally reached out to Isabelle Roumeguere in order to confirm the existence of the bow and arrow but also to get clarification on the nature of X.

I am very much indebted to Christophe Kilian here who provided me with a valuable article on Isabelle Roumeguere which helped me track her down. While on the subject of Christophe, we also discussed the Yeti situation in the Pyrenees where I hope to join at some point in the near future.

I'm still in the process of narrowing down a more specific territory, especially for the expedition I'm currently planning in Kenya with cultural anthropologist Kenneth Joholske and archeologist Angelique Botes-Guthrie.

THE GHOST SPECIES

We now know thanks to genetics that modern *Homo sapiens* were born 300,000 years ago in Sub-Saharan Africa, including the region that X is reported from. We also know that between 300,000 years ago to sometime under 100,000 years ago in Africa, and at least up to 25,000 years ago in Eurasia multiple Hominin species co-existed and shared genes with each other, including modern humans.

Many of these now officially extinct human relatives resembled the various X reported currently in Africa. We also know from genetics that Sub-Saharan Africa today is not only the oldest genetics on earth, but also has the most amount of diversity. Think about this. Every element is present here for the birth and subsequent survival of a species such as X, even maybe more then one type? The question is no longer whether is X possible? It is, it was. The question is, is there now and what do we do about it?

So, are X1 and X2 relic Hominins from sometime between 300,000 years ago and let's say 50,000 years ago and a result of mixed genetics between *Homo sapiens* and more primitive relatives? I don't know, and we can easily get into unfavorable speculation here, so we have to trend with respect for indigenous cultures, but with science and genetics as our tools to not only further expand our knowledge of our world and history but also with the intent of giving back to African culture and history because that is also all of our history as well.

As a quick last note geneticists have also found in our genes a 'Ghost species' apart from Neandertals and Denisovans. This 'Ghost Species' is an ancient primitive relatives that interbred with our own specie, in you guessed it Sub-Saharan Africa. This is just the beginning and with new tools like Environmental DNA, the future of our past is so very bright.

Marine monsters on Spanish beaches

By Javier Resines

Every once in a while, Spanish cryptozoology enthusiasts are given the news of the appearance of strange animals stranded on our beaches. When the remains are found are in good condition, there is usually no doubt as to the identification of the species. The problem is complicated, however, as the symptoms of decomposition begin to become visible and time and the elements make a dent in the body of the poor creature.

When a good part of the body mass disappears, its identification with the naked eye is difficult; it may even appear that we are facing a true monster, a creature out of our worst nightmares, that - suddenly - has arrived on the beach of our city.

Let`s be cautious. The cases of stranding of animals of considerable size are many, but the really unidentified specimens are reduced to a small percentage. Let's look at an example. Between 2000 and 2015, a total of 235 stranded cetaceans were found in the Doñana Natural Park, a small coastal wetland located southwest of Spain.

Of these, almost half corresponded to dolphins, followed by porpoises, whales, sperm whales and whales. Strandings of species such as the Blainville zifio (beaked whale) (*Mesoplodon densirostris*) and the Cuvier's zifio (*Ziphius cavirostris*) were also detected. These last two are especially significant because they are little known species and possess a particular anatomy that, when suffering the rigors of decomposition, can reveal suggestive morphologies similar to the remains of legendary sea monsters.

But, of the 235 animals stranded in those years, 38 have not been identified. This circumstance does not mean that we are faced with new marine species unknown to science, but that - simply and for whatever reasons - they have simply *not* been identified. Lack of means, of knowledge, of time; there are many reasons that are offered to us although - among them - there could be the one that really interests us: to find ourselves in front of a new creature for the biology books.

Recently on our coasts we have had examples of strandings that have been effectively studied and have managed to determine which species they belong to, but there are others which continue to sow doubt as to whether or not, on this occasion, we are facing an unknown creature.

Strange animal on a beach in Islantilla, Huelva.

. Photo: Jose Cabello.

On March 7, 2019, a neighbour found the remains of what looked like a strange marine animal on the Islantilla beach in Huelva, and concluded that it had been possibly dragged to the shore as a result of the storm that affected the area in recent days. After the initial surprise, and unable to determine which species the body might belong to, the witness decided to photograph it and send the images to a nearby cultural institution, the House of Science of Seville.

Comparison between the remains of Islantilla and a basking shark

Animal stranded in Islantilla. Photo: CanalCosta TV

Creature of Villarico

This institution contacted ichthyologists and marine biologists of the CSIC (Higher Center for Scientific Research, an official scientific body) who could not identify which animal it was. They only managed to certify that the remains belonged to a

"The monster of Ibiza" in a capture of the video edited by Unknown Mystery group.

vertebrate, with a tail and cartilage in the area that seemed to be its head.

Due to the state of decomposition in which the remains were found, they were removed from the beach by a crane belonging to the Maintenance Service of the Commonwealth of Islantilla, before a sample of the tissues could be taken for later analysis. Three or four days later, the House of Science contacted the Commonwealth of Islantilla to find out where the body was buried. They were aware that they had lost a great opportunity to solve the mystery.

Thus, researchers from the Institute of Marine Sciences of Andalusia (ICMAN) moved to the site to take DNA samples of this unique creature. Samples were sent quickly to the University of Vigo, where Rafael Bañón, one of the world's leading experts in the identification of marine species, works.

He asked to have six samples of the body sent to him so that he could carry out a DNA test. Together with their usual team, they discovered that they were facing the remains of a basking shark, a species that lives in winter in the Gulf of Cádiz and - in spring - they begin their migration to northern Europe in search of the plankton banks on which they feed.

"For us, performing a DNA test of this type is routine, but we were very excited to be able to participate in this work, because we helped solve a mystery," said biologist Ángel Comesaña, an expert who also participated in the study.

In this case, the mystery was solved thanks to having preserved the remains of the animal for later analysis, a practice that should be basic in cases of stranding of potentially strange creatures but that does not always happen. On this occasion, the monster was not such a monster. But it's not always so clear …

In mid-August 2013, the beach of Villarico (Almería), also on the Andalusian coast, witnessed a similar event. An unusual animal of more than four meters in length was found on the shore, also in an advanced state of decomposition. The dead creature caused astonishment amongst the bathers gathered there, who promptly notified the emergency service. Civil Protection volunteers extracted the remains of the creature from the water, where it had been entangled in a rope or something similar They cordoned off the area and requested the participation of the Nature Protection Service (Seprona) and the Marine Wildlife Defense Association (Promar) of the neighboring municipality of Pulpi.

"In summer we are watching the beaches; a lady found a part and we already helped take out the rest. We have no idea what it can be; smell, it smelled bad, because it was very broken. Promar experts are seeing what could be," explained the Civil Protection coordinator of the area.

Indeed, Promar indicated that - due to the advanced state of decomposition of the marine animal - it had been impossible to identify it, so a series of photographs were been sent to experts belonging to other entities associated with the stranding network that can determine the species to which it belongs.

Although it seems to be ruled out that it could be some type of mammal and that would be a fish, it will not be possible to determine it at the moment with DNA tests, because no entity has been responsible for the expenses that this analysis would entail.

Therefore, the remains of the animal have been buried - paying attention to health reasons - after being studied by Seprona and Promar. This last association, rejected the possibility of conserving the remains for study because it is not presumably a mammal, its main research area ...

It is a great pity that the lack of interest and budget deprives us of an interesting study about something we will never be sure of what it was. An oarfish? A basking shark, as in the case of Islantilla? An unknown animal? That said: we will never know ...

In Spain there are dozens of cases similar to that of Villarico in which the stranded animal was never analyzed and, therefore, never let itself be known if we are facing a new species. An example of this is the case known as the "monster of Ibiza", an event studied by the Spanish team of researchers of *Mystery Unknown*. On March 14, 2014, a group of amateur fishermen found the remains of an animal - in an advanced state of rot - in an area of rocks located on a beach especially appreciated for its tourist interest, located in the town of Sant Antoni, in the island of Ibiza.

Apparently, they were facing some serpentiform remains, basically formed by fat and skin debris, attached to a spine, which prevents identifying the animal in question. The appearance is that of a true sea monster but, in the absence of evidence, we can only speculate with this possibility.

Because, as I warned at the beginning of the article, we can only be prudent. Surely, all (or almost all) the remains found belong to known species. It is in that "almost" where we house the small (by chance) but great illusion of discovering that - for once - the news of a stranger or a strange creature is not going to be forgotten.

Cuvier's beaked whale skeleton stranded in Fuerteventura (Canary Island)in 2004 and exposed on Los Cotillos beach.

ENTER AFONYA: THE MOST DETAILED WILDMAN ENCOUNTER OF ALL

Richard Freeman

In the former USSR the most famous of the wildmen was known as the almasty. These man-like beasts were reported from the Caucasus, Altai and Pamir Mountains. Ranging from human-sized to around eight feet tall, the almasty is covered with hair except for the face. It has a thick brow-ridge, sloping brow, flat, wide nose, a wide mouth with thin lips and a muscular build. It does not use fire but can hurl rocks and swing clubs. Omnivorous, it's ecological twin is the brown bear.

Reports of the almasty have been made close to the borders of Europe. One of the most detailed, multi-witness cases happened in 1988 on the Kola Peninsula.

In the summer of 1987, a group of six youths aged between 15-18 from the town of Lovozero fished, and picked berries and mushrooms around a lake of the same name on the Kola peninsula in western Russia close to the border with Finland. The boys had built a wooden cabin. The building was raised up on fir stumps as the area was prone to flooding in springtime. On August 11th the boys were sitting around a campfire. They had the odd feeling they were being watched. Five of them retired to the shack. One of them Sasha Prikhodchenko, lay down to sleep by the campfire. As he peered out from under his blanket, he saw a pair of huge, hair-covered legs approaching the shack from behind. The legs were visible due to the shack being raised on tree stumps. Sasha ran back to the shack and told his friends. Looking out they saw a huge, man-like creature covered in gray hair. It circled the shack. Terrified, the boys used a stick to bolt the door and stayed awake all night. They nicknamed the monster 'Afonya'.

In the light of day, the boys became braver. Nothing happened all day but as night fell, they decided to scare Afonya off with noise. They played pop music full volume on a tape recorder and threw rocks into the bushes. As they sat around the fire a rock flew out of the bushes and landed in the fire. Shortly after another struck the cabin. The group took shelter in the cabin as it was bombarded with rocks. One of them looked out of the door whilst protecting his head with a metal pot. He saw the almasty moving back into the forest. Later, Afonya returned and beat the outside of the cabin. The boys reasoned that the creature had been sleeping in the cabin in their absence and was now trying to drive them out.

Next day they secured the door with a stick and returned to the town on their motorboat. They told other people of their adventure but were roundly disbelieved and laughed at. They returned to the cabin and found the stick still in place. That night Afonya failed to appear. One boy suggested that they tried to surround the almasty and kill it with axes but the others feared that the monster would kill them all. Instead they decided to try and befriend the beast.

Next day they left a table of food out for Afonya and went back to their boat. Later they saw the creature on the shore. He seemed alarmed by their outboard motors and vanished. The boys returned and found the food untouched. They decided to stay in the cabin again.

Before going to sleep one of the group, Slava Kovalev stepped outside of the cabin to urinate. As he finished, he looked up to see the almasty only six feet from him. Slava ran into the cabin and bolted the door with a stick. Afonya came up to the door and pushed it, snapping the stick like a matchstick. As the door swung open the monster stood in the doorway. The panicking boys scrambled under the beds but Afonya did not enter but simply banged on the door. The frightened boys dare not approach the door to close it. They stayed awake all night and heard the almasty leap onto the roof and walk about. As the sun rose, he left.

The following day three more boys arrived from the town and the original six were emboldened to stay. As they sat around the fire that night Afonya appeared. He was squatting in an ape-like stance on all fours some forty- five feet away. He bounded back and forth on all fours then charged. The boys scrambled into the shack. Slava held the door and shouted for help. Roman Leonov tried to push an axe handle through the metal hoops that had held the now smashed stick bolt. As he did so the monster gripped the half- opened door. Roman got a close look at Afonya's hand which he described as like a man's only much bigger. It was sparsely covered in gray hair and had dark skin.

The boys frantically pulled at the door as they tried to slide the axe handle through the bolts. The almasty pulled back and ripped the door open. Roman got another good look at the beast, this time his face.

"I saw his face right in front of mine. Mind you, he was standing on the ground and I was on the floor which is raised high above the ground. His face is brown and wrinkled. Somehow, I didn't notice his nose and mouth' I just remember they eyes glittering, angry and reddish. They are set far apart, like a horse's, sort of looking sideways. The forehead is wide and polished, the head is round like a ball. I can't remember how I got under the plank-bed and dropped the axe. I came back to my senses because of a strong knocking on the walls and the boys yelling like mad. The knock was so strong that I dashed back to the door. It was open and I saw Slava Kovalev by the fire and moaning from pain. As I learned later, when Afonya pulled the door sharply, Slava was thrown out of the cabin and hit his shoulder against the door-frame. He was so angry when he returned to the cabin, he said he would go out and face our adversary, but we asked him to stay inside the cabin."

Afonya leapt onto the roof again and stomped around so loudly the boys feared he would crash through the roof. The monster apparently burnt his hand on the hot chimney that was attached to the burning stove. The almasty made a mooing sound, jumped down and retreated into the forest.

In the morning six boys returned to town and three remained at the cabin. They spent the day fishing. On returning to the cabin they saw Afonya approaching quickly. The boys climbed onto the roof via a ladder and cringed in fear but the monster simply turned and went back into the forest. As soon as he had gone the boys ran for their boat and motored back to the town.

Back in town the boy's reports were treated with scorn once more. They were even accused of taking drugs.

On August 18th the group returned with three more boys from the town. Now twelve boys were searching for the monster. After seeing Afonya prowling the shore from their boats the three newcomers chickened out and turned their boat around. The almasty hurled rocks at the boats as well as a branch from a birch tree that almost struck one of the boats.

They returned the next day and fished from their shore except for Sasha Prikhodchenko who was cleaning the table outside of the cabin. The boys in the boats saw Afonya approaching and shouted a warning to Sasha who thought they were joking as on all previous occasions the creature had only appeared at night. But looking up he saw the creature only fifteen feet from him. Sasha ran for the cabin and bolted the door with an axe handle. The others ran for their boats.

The boys returned later to check on Sasha and saw the almasty on the shore. He kept pace with their boats and did not let them come to shore. When he eventually vanished into the trees, they cut the motors and rowed to shore. Three boys Slava Kovalev, Sasha Sveilis and Roman Leonov decided to land. As they walked towards the cabin Afonya leapt from the bushes and chased them. Slava and Roman jumped into their boat and started the motor. Sasha's boat was stuck and as he struggled with it the motor fell into the water. Afonya came up behind him and the other boys shouted to him to jump into their boat. Sasha, however, was paralyzed with fear.

The almasty crouched then stood, watching the boy. The others rowed alongside and told Sasha to climb into their boat. As he tried to do so he glanced back at the monster and fainted. The other boys pulled him into their boat and splashed water on him to awaken him. Rowing out again, they waited till the beast had retreated.

Sasha Prikhodchenko, who had been watching from the cabin,shouted out that the creature had retreated beyond a creek. Arming themselves with axes, rocks and hammers the others returned to try and rescue Sasha from the cabin.

Afonya ran out of the forest again, right into the group, cutting it in two. One of them, Slava Surodin hurled a rock at him. It struck Afonya in the shoulder and he made a mooing sound. Two of the group ran for the cabin the others for the boats. The almasty chased the boys who ran to the cabin.

One of them, Ivan Dyba said …

"As I felt him literally breathing down the back of my neck, my legs failed me. I snatched at Zhenya Trofimov running in front of me and we both fell to the ground. We crawled and scrambled into the cabin."

The terrified boys hid inside as the monster pounded the walls and peered through the window. Yet again the almasty retreated into the forest and the other boys tried to rescue their comrades. As if he were playing cat and mouse with them Afonya charged out of the undergrowth again. Sasha Prikhodchenko and Zhenya Trofimov ran back to the cabin whilst Ivan Dyba leapt into the water fully clothed and swam to a boat. Those on the boat pulled him aboard and motored to the opposite shore to dry his clothes.

Deciding to alert the authorities the three boys on the boat returned to town. Arriving at 2 am they went to the town's executive committee. The woman on duty called a patrolling militiaman. After listening to their story he called a senior game warden called Kuznetsov who arrived an hour later. Laughing at the boy's story, he said ...

"I am not going to alert anyone in the dead of night. I don't want to become a laughing stock on account of your 'snowman'. I'll send someone over in the daytime."

On August 20th, help came to the boys trapped in the cabin in the form of a game warden called Igor Pavlov and his two assistants. All had rifles After listening to the story they arrived at 11 pm. The summer that far north still giving plenty of light the boys alerted the men when Afonya emerged. Igor had been told by his superiors that the boys had been frightened by an 'old bear' but what he saw some 82 feet away looked more man-like but Pavlov saw it was clearly not human. The monster ran back into the forest but the warden was able to gauge the creature's height against a branch sticking out from a tree. Afonya was at least 8 feet tall. He noted that the almasty's arms reached it's knees and it was covered in gray hair. It ran with long strides.

The adults and boys stayed in the cabin that night. The building was bombarded with rocks. In the morning the boys left for home but the wardens stayed on. They found a number of tracks imprinted more deeply than an adult man. They spent another night at the cabin but were not disturbed.

The next day, several adults accompanied the boys to the site with torches and cameras. Afonya hurled rocks at the cabin but did not show himself. Soon after the area was over run with film crews and reporters. Afonya by this time had vanished.

The location of these events is close to the border with Finland. Scandinavia is well known for its tradition of trolls, shambling, hairy, club-wielding, man-eating giants of ancient legend. Lars Thomas of Copenhagen University was studying ancient texts pertaining to a legendary Danish king who loved to hunt. His favorite quarry were trolls because of their savagery when cornered. The descriptions of trolls in these writings spoke of tall, muscular, hair-covered creatures much like men. They had thick brow ridges, long arms, deep set eyes and the females had long breasts . Trolls had no fire but could hurl large rocks and use clubs. The description matched up uncannily well with modern day descriptions of the almasty that I had recently collected from witnesses in the Caucasus Mountains in Russia. It is likely that these creatures, on occasion, cross over into Europe and seeded the stories of trolls in Scandinavia.

The description of the almasty seems more man-like that the larger yeti, yeren, mande-barung and sasquatch. There are even stories of it cross-breeding with modern humans, though this may well be folkloric. The Russian wildman may be a very large, robust, early offshoot of *Homo erectus.* One adapted to mountainous forests. Instead of going down the road of complex tool use and fire making it utilized it's massive strength. Alternatively it could be that the creature descended from Homo habilis, as we now know this hominin, that we once thought had never left Africa, had close relations in tropical Asia half a world away. It must have left other offshoots on it's long migration. Finally there is the possibility that the almasty is of an utterly unknown hominin species. We know from genetic markers on populations of modern humans show that their ancestors mated with utterly unknown hominin species for which we do not even have fossil remains. Only further field work will solve the riddle of the almasty. The situation at the time of writing looks like further expeditions in Russia will not be possible for some time. I would suggest concentrating on Central Asia and countries like Tajikistan and Kyrgyzstan for the best results.

The Centre for Fortean Zoology,
Myrtle Cottage,
Woolfardisworthy,
Bideford, North Devon
EX39 5QR

Telephone 01237 431413
Fax+44 (0)7006-074-925
jon@eclipse.co.uk

www.cfz.org.uk

THE CENTRE FOR FORTEAN ZOOLOGY
Annual Report 2021

Dear friends,

I cannot believe that for the 27th time I have sat down to dictate the Centre for Fortean Zoology Annual Report. In fact, I'm not writing it, I am dictating it to my amanuensis Louis, because my hands are not working particularly well at the moment, but enough of my bellyaching. Whilst, I suspect in common with most organisations working in a similar situation to us, this year has been disappointing because of the strictures caused by the worldwide Coronavirus pandemic, some things have been done and once again it is Louis who is responsible for most of it.

All the expeditions and UK based investigations that we had planned for this year ended up being cancelled, that is - except for - the investigation into lynx-sized cats in the Forest of Dean which has been run by Carl Marshall since the end of 2019. For those of you who are not aware of what, I believe in the current vernacular is called the backstory, Carl was in the Forest of Dean with a friend of his and his young son in the closing months of 2019. They were looking for wild boar, which indeed, they found, but they also found a footprint that gives every indication of being that of a felid far larger than any domestic cat, although smaller than that of a Puma or Panther.

He sent photographs of this footprint to a number of well-respected zoologists, who with only one exception, agreed with him. Since then we have been running a series of investigations in the Forest of Dean to find out the likelihood of there being Lynx or similar sized cats in the forest. Here it should be noted that the Forest of Dean is not particularly far away from the parts of Shropshire where Dr Karl Shuker has established that not only were there jungle cats (*Felis chaus) living there at one time reasonably recently, but has also indicated the likelihood that genetic material of this species have entered the gene pool of the area's feral cats. This indicates that the idea of a medium sized cat species living wild in the area is not at all unlikely.*

However, the intensive trail camming which we have carried out has produced the expected result of showing that there are numerous deer and rodents living in the location. What appeared to have been a den for one of our target carnivores appears either to have been deserted, or nothing of the sort. The only hair samples that we have secured, to date, have proven not to be from a cat species. However, the investigation will continue in the New Year and I hope that in twelve months time I will have more positive news for you!

On a positive note, the requirement for camera trapping in the Forest of Dean has triggered Louis Rozier to start work on designing a CFZ trail cam, using recycled smartphones. These devices will allow us to affordably deploy cameras to sites of interest and will return photos directly to us without need for constant tending to, which could make it significantly easier for us to conduct these sorts of investigations in the future. Louis has already produced and tested a basic working prototype and we will look to develop this concept further in the new year.

Louis's biggest contribution during the last twelve months, however, is the beautiful new website that he designed and developed for us, which went live for the first time back in the spring. It supersedes both the old website, and the CFZ blogs which have been running on the Google owned Blogger since 2005. The old blogs are still there as legacy sites, and can be reached through the new current site. Louis is also busy at work developing some new web software for communities for one of his own projects, or "side-hustles" as the kids call it nowadays, and we may look to integrate this into our site in the future.

As I am sure you are all aware, my beloved wife Corinna died of cancer during the summer of 2020 and as I am sure you can imagine, recovering from such a serious blow has not been easy. Indeed I am not quite there yet but we are getting there slowly. But we are still not functioning at the level that we would like to be working at.

This year we managed to publish two books, both of which we had been working upon for a considerable length of time. The first is 'The Soviet Sasquatch' by Boris Porshnev:

In 1957 Soviet historian and social scientist Boris Porshnev, inspired by the reports of the Yeti in the Himalayas, became interested in the possibility of similar creatures in the area of Europe and Asia then controlled by the Soviet Union. He was given permission by the Soviet Academy of Sciences to establish a Commission to examine the whole question of the 'Snowman'. After he wrote an article in Pravda he received over a thousand reports from all over the Soviet Union, giving a consistent picture of a wild creature, more closely related to human beings than any known species, surviving in mountainous areas all over Asia. An expedition to the Pamirs of Tajikistan was organised in 1958 to follow up the most promising reports. Unfortunately, more powerful figures in the scientific establishment subverted the original purpose of the expedition and it produced little result. From then on Porshnev's position declined. His theory that Asian wildman reports could be explained by surviving Neanderthals was attacked, in one case in terms that doubted his sanity. The defence he wrote could not be published in Moscow, and had to appear in a Kazakhstan literary magazine.

In 1963 he produced a book summarising the evidence the Commission had received, studies from other parts of the world, and further evidence from history. He built up on this basis a consistent picture of the creature and discussed its possible relation to Neanderthal man. The book was never actually completed, but 180 copies of a preliminary version were circulated to colleagues in Moscow. The book then disappeared for well over half a century. With the assistance of Porshnev's family the manuscript has now reached the West and is published here in an English translation with the addition of notes, maps, illustrations and an index. This book casts a wholly new light on the Yeti, Bigfoot and the possible survival of human ancestors into the present day.

The second being my book about my childhood in Hong Kong: Wild Colonial Boy

Anybody who has been a CFZ watcher at any time this last thirty years, will know that I have been promising to write a book about my childhood in Hong Kong, and my early introductions to the arcane world of Fortean zoology. Well, half a century after I first thought of the idea, and over forty years since I actually started writing the bloody thing, the first edition is finally out.

For those of you who are interested in the more Fortean aspects of the stories, let me assure you it includes a bunch of musings on the subject of the final Hong Kong tiger, what happened to Hong Kong's foxes, accounts of mysterious apes in the heavily forested areas on the south of the island, the complicated story of St. John's macaque, rumours of giant earthworms, and all sorts of other things besides. I stress that it is the first edition, because there are still a few minor typographical errors needing to be sorted out. Louis and I are working on them as we speak, and so the first edition – especially one signed by me – is likely to become a collector's item in very short order.

I have had several people write to me, having brought my book, wondering why I have been so blunt about my relationship with my parents over the years. The fact that this book was published in the same year that the Duke of Sussex and his unfortunate wife made a whole bunch of revelations and claims about the Royal Family that is difficult to see as anything other than cheap points scoring is unfortunate. However, my intentions in writing this were far more honourable. Since my seismic life change in August 2000, I have had to reevaluate my life, personal universe and everything once again. The last time I had to do this on such a level, was when I split up with my first wife Alison, in the summer of 1996. This time I will not be using as my main treatment tools alcohol and narcotics, I am too old and I fear too sensible these days, so I have finally finished the book I have been working on for decades, in which I attempt to address - amongst other things - various things that Roger Waters would have described as "bricks" in my own personal wall. I loved both of my parents and in fact still do, despite the fact that my mother has been dead twenty years and my father fifteen, but my relationship with them was neither good nor easy and was rather dysfunctional most of the time. In this book I spend most of my time talking about natural history and cryptozoology, and the life of a young boy in Hong Kong in the 1950's, but I have not shied away from the things that Phillip Larkin talks about in one of his most famous poems 'This be the verse'. At least in my case, my mum and dad did indeed fuck me up.

However, I didn't write it in order to point blame at them, nor would I want anybody to read it in that way. I wrote it to try and make sense of my childhood and to try to offer an answer to people who have asked me what my childhood was like.

Now for some bad news, both Martha the pigeon that Corinna hand reared and her pet crow Bard died this year. Martha was near the top of the age limit for captive pigeons and Bard had always been sickly. We miss them both, and this means that at the moment, not counting Archie and the cats, the CFZ menagerie consists of two chickens, two axolotls, an african leaf fish, Clarence the clarias Catfish, the tank in my sitting room which has a few mosquito fish and three paradise fish, which are showing the beginning of signs of wanting to breed. I am not intending to get any more animals any time soon, but in the past whenever I have said that the universe has proven my resolution to be wrong.

We are currently working on the 2022 yearbook, being compiled by me and Richard Muirhead, I hope it will be out by the Spring. We are also working on the long delayed new issues of Animals and Men, both of these things are very difficult to do without Corinna who was as responsible for them as I was over the past fifteen years.

The most important thing that we have done over the past year concerns our weekly webTV show, which has been ongoing since 2007. However, it only became weekly in the Spring of this year. Again, I want to thank Louis for everything he has done to help us improve the production value and streamline the production schedule for the show. Louis, my friend, I cannot thank you enough. On The Track is the thing I enjoy most, and I am very happy to bring it to you each week on Saturday on CFZTV.

As we approach the new year, there is a lot to look forward to. Back in the summer Richard and I bought a stack of papers from the late Odette Tchernine, including an unpublished manuscript which we will be publishing ourselves during 2022. However, when we looked at the collection of press cuttings, letters and unpublished writings, we decided that it behoves us to celebrate the legacy of this remarkable lady. CFZ volunteer Guin Palmer has taken on the mantle of collecting more of Odette's writings and memorabilia and together with Louis, will be assembling an in depth website for and about the lady. Guin has done a remarkable job and we look forward very much to sharing the results with you all.

Also, in 2022 we have the first book by Damon Corrie, an Amerindian Chieftain of the Arawak tribe in Guiana in which he describes - for the first time - many of the animal archetypes of his people.

As regular viewers of On The Track will know, we are in the early stages of setting up a, or at least carrying out a feasibility study, into a research project on the Japanese island of Iriomote. As you probably know, this island, one of the southernmost parts of the Japanese archipelago is home to a unique species of cat *Prionailurus iriomotensis* which some people believe is a subspecies of the Asian Leopard Cat. It was discovered in 1965 by Tetsuo Koura, who also looked into rumours of a larger mystery cat on the island. This, if it exists, will almost certainly turn out to be a clouded leopard of some description, but the important thing is to find out to which of the two species of clouded leopards to which it belongs. However, as recent evidence has shown that speciation within the clouded leopard complex is not unknown, could it be a new subspecies, and if so, what is its relationship to the semi-mythical clouded leopards of Formosa.

We are looking for some volunteers to help put this project together. This, in the early stages, will take completely on social media, but if there is anybody who speaks Japanese or who reads this and has family or friends on Iriomote and wants to have a bash at getting themselves a modicum of zoological immortality, please drop me a line at:

CFZJon@gmail.com

...and finally:

Thank you very much to all of you who have stuck with me through the past few years, many of you had noticed that both the CFZ and I had taken, to a certain extent at least, a backseat as far as Cryptozoology was concerned. You all know now that it was because of the illnesses of Corinna and Mother, both of them were dying and under the circumstances obeying Corinna's wishes and not telling anybody was not only the kindest thing that I could do, but the only thing that I could do for her. However, as I'm sure you can understand, dealing with the aftermath of the death of these two people whom I loved and love very much, has not been easy, but I am coming out of the other side and would like to think that the CFZ is back on an upward trajectory.

I would like to thank everybody who works so hard for the CFZ, and also the people who – each month – volunteer to join our merry band of brothers and sisters. If you want to be one of these gallant volunteers, please email me at cfzjon@gmail.com.

Many good wishes to you all from me for a Happy Christmas and a peaceful and fulfilling New Year for you and yours!

Yours, as ever,

Jon Downes
(Director, Centre for Fortean Zoology)
24.12.21

STILL ON THE TRACK OF UNKNOWN ANIMALS

The Centre for Fortean Zoology, or CFZ, is a non profit-making organisation founded in 1992 with the aim of being a clearing house for information, and coordinating research into mystery animals around the world.

We also study out of place animals, rare and aberrant animal behaviour, and Zooform Phenomena; little-understood "things" that appear to be animals, but which are in fact nothing of the sort, and not even alive (at least in the way we understand the term).

Not only are we the biggest organisation of our type in the world, but - or so we like to think - we are the best. We are certainly the only truly global cryptozoological research organisation, and we carry out our investigations using a strictly scientific set of guidelines. We are expanding all the time and looking to recruit new members to help us in our research into mysterious animals and strange creatures across the globe.

Why should you join us? Because, if you are genuinely interested in trying to solve the last great mysteries of Mother Nature, there is nobody better than us with whom to do it.

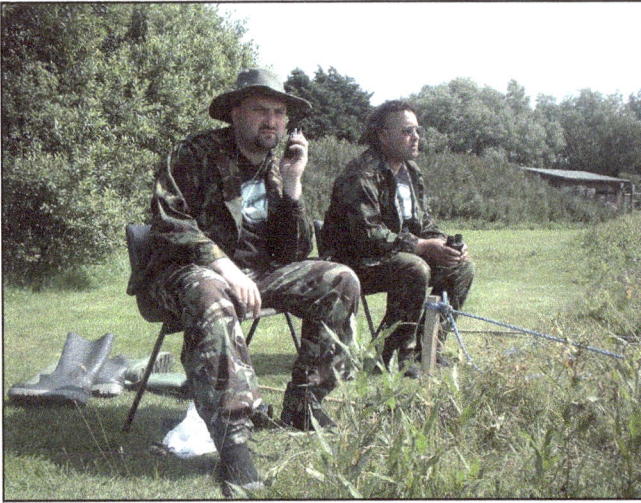

We publish a journal *Animals & Men.* Each issue contains nearly 100 pages packed with news, articles, letters, research papers, field reports, and even a gossip column! The magazine is Royal Octavo in format with a full colour cover. You also have access to one of the world's largest collections of resource material dealing with cryptozoology and allied disciplines, and people from the CFZ membership regularly take part in fieldwork and expeditions around the world.

The CFZ is managed by a board of trustees, with a non-profit making trust registered with HM Government Stamp Office. The board of trustees is supported by a Permanent Directorate of full and part-time staff, and advised by a Consultancy Board of specialists - many of whom are world-renowned experts in their particular field. We have regional representatives across the UK, the USA, and many other parts of the world, and are affiliated with other organisations whose aims and protocols mirror our own.

You'll find that the people at the CFZ are friendly and approachable. We have a thriving forum on the website which is the hub of an ever-growing electronic community. You will soon find your feet. Many members of the CFZ Permanent Directorate started off as ordinary members, and now work full-time chasing monsters around the world.

Write to us, e-mail us, or telephone us. The list of future projects on the website is not exhaustive. If you have a good idea for an investigation, please tell us. We may well be able to help.

We are always looking for volunteers to join us. If you see a project that interests you, do not hesitate to get in touch with us. Under certain circumstances we can help provide funding for your trip. If you look on the future projects section of the website, you can see some of the projects that we have pencilled in for the next few years.

In 2003 and 2004 we sent three-man expeditions to Sumatra looking for Orang-Pendek - a semi-legendary bipedal ape. The same three went to Mongolia in 2005. All three members started off merely subscribers to the CFZ magazine. Next time it could be you!

We have no magic sources of income. All our funds come from donations, membership fees, and sales of our publications and merchandise. We are always looking for corporate sponsorship, and other sources of revenue. If you have any ideas for fund-raising please let us know. However, unlike other cryptozoological organisations in the past, we do not live in an intellectual ivory tower. We are not afraid to get our hands dirty, and furthermore we are not one of those organisations where the membership have to raise money so that a privileged few can go on expensive foreign trips. Our research teams, both in the UK and abroad, consist of a mixture of experienced and inexperienced personnel. We are truly a community, and work on the premise that the benefits of CFZ membership are open to all.

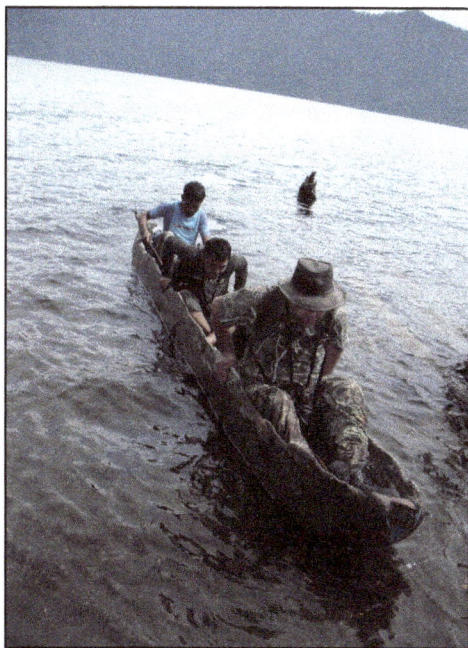

Reports of our investigations are published on our website as soon as they are available. Preliminary reports are posted within days of the project finishing.

Each year we publish a 200 page yearbook containing research papers and expedition reports too long to be printed in the journal. We freely circulate our information to anybody who asks for it.

We have a thriving YouTube channel, CFZtv, which has well over two hundred self-made documentaries, lecture appearances, and episodes of our monthly webTV show. We have a daily online magazine, which has over a million hits each year.

From 2000—2016 we held our annual convention - the Weird Weekend. It went on hiatus because of the illness of several of the major personnel and the eventual death of one of them. But we plan to bring it back soon. It is three days of lectures, workshops, and excursions. But most importantly it is a chance for members of the CFZ to meet each other, and to talk with the members of the permanent directorate in a relaxed and informal setting and preferably with a pint of beer in one hand. Since 2006 - the Weird Weekend has been bigger and better and held in the idyllic rural location of Woolsery in North Devon.

Since relocating to North Devon in 2005 we have become ever more closely involved with other community organisations, and we hope that this trend will continue. We have also worked closely with Police Forces across the UK as consultants for animal mutilation cases, and we intend to forge closer links with the coastguard and other community services. We want to work closely with those who regularly travel into the Bristol Channel, so that if the recent trend of exotic animal visitors to our coastal waters continues, we can be out there as soon as possible.

Apart from having been the only Fortean Zoological organisation in the world to have consistently published material on all aspects of the subject for over a decade, we have achieved the following concrete results:

• Disproved the myth relating to the headless so-called sea-serpent carcass of Durgan beach in Cornwall 1975

• Disproved the story of the 1988 puma skull of Lustleigh Cleave

- Carried out the only in-depth research ever into the mythos of the Cornish Owlman.
- Made the first records of a tropical species of lamprey
- Made the first records of a luminous cave gnat larva in Thailand
- Discovered a possible new species of British mammal - the beech marten
- In 1994-6 carried out the first archival fortean zoological survey of Hong Kong
- In the year 2000, CFZ theories were confirmed when a new species of lizard was added to the British List
- Identified the monster of Martin Mere in Lancashire as a giant wels catfish
- Expanded the known range of Armitage's skink in the Gambia by 80%
- Obtained photographic evidence of the remains of Europe's largest known pike
- Carried out the first ever in-depth study of the ninki-nanka
- Carried out the first attempt to breed Puerto Rican cave snails in captivity
- Were the first European explorers to visit the `lost valley` in Sumatra
- Published the first ever evidence for a new tribe of pygmies in Guyana
- Published the first evidence for a new species of caiman in Guyana
- Filmed unknown creatures on a monster-haunted lake in Ireland for the first time

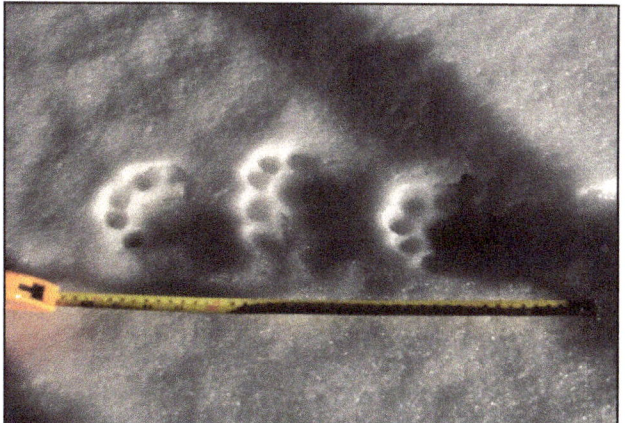

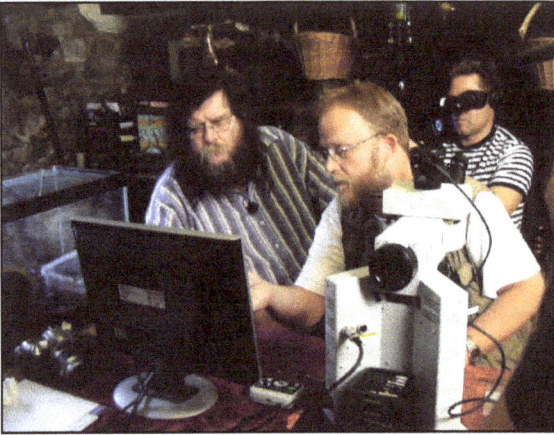

- Had a sighting of orang pendek in Sumatra in 2009
- Found leopard hair, subsequently identified by DNA analysis, from rural North Devon in 2010
- Brought back hairs which appear to be from an unknown primate in Sumatra
- Published some of the best evidence ever for the almasty in southern Russia

CFZ Expeditions and Investigations include:

- 1998 Puerto Rico, Florida, Mexico (Chupacabras)
- 1999 Nevada (Bigfoot)
- 2000 Thailand (Naga)
- 2002 Martin Mere (Giant catfish)
- 2002 Cleveland (Wallaby mutilation)
- 2003 Bolam Lake (BHM Reports)

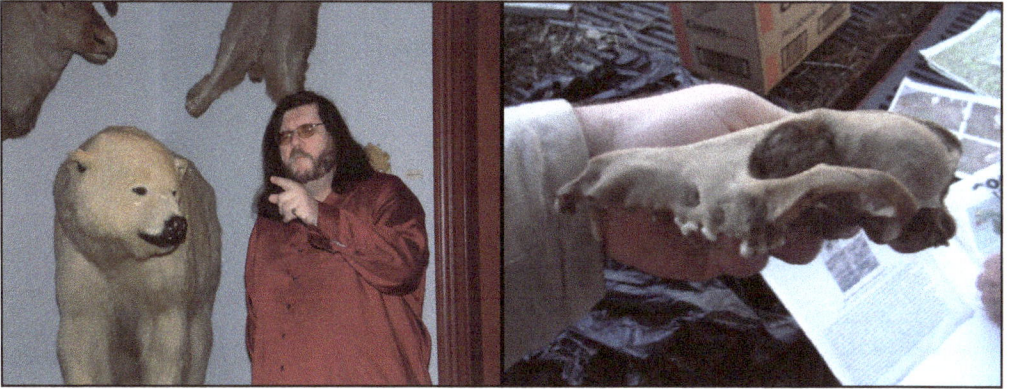

- 2003 Sumatra (Orang Pendek)
- 2003 Texas (Bigfoot; giant snapping turtles)
- 2004 Sumatra (Orang Pendek; cigau, a sabre-toothed cat)
- 2004 Illinois (Black panthers; cicada swarm)
- 2004 Texas (Mystery blue dog)
- Loch Morar (Monster)
- 2004 Puerto Rico (Chupacabras; carnivorous cave snails)
- 2005 Belize (Affiliate expedition for hairy dwarfs)
- 2005 Loch Ness (Monster)
- 2005 Mongolia (Allghoi Khorkhoi aka Mongolian death worm)

- 2006 Gambia (Gambo - Gambian sea monster , Ninki Nanka and Armitage's skink
- 2006 Llangorse Lake (Giant pike, giant eels)
- 2006 Windermere (Giant eels)
- 2007 Coniston Water (Giant eels)
- 2007 Guyana (Giant anaconda, didi, water tiger)
- 2008 Russia (Almasty)
- 2009 Sumatra (Orang pendek)
- 2009 Republic of Ireland (Lake Monster)
- 2010 Texas (Blue Dogs)
- 2010 India (Mande Burung)
- 2011 Sumatra (Orang-pendek)
- 2012 Sumatra (Orang Pendek)
- 2014 Tasmania (Thylacine)
- 2015 Tasmania (Thylacine)
- 2016 Tasmania (Thylacine)
- 2017 Tasmania (Thylacine)
- 2018 Tajikistan (Gul)
- 2020 Forest of Dean (Lynx)

For details of current membership fees, current expeditions and investigations, and voluntary posts within the CFZ that need your help, please do not hesitate to contact us.

The Centre for Fortean Zoology,
Myrtle Cottage,
Woolfardisworthy,
Bideford, North Devon
EX39 5QR

Telephone 01237 431413
Fax+44 (0)7006-074-925
eMail info@cfz.org.uk

Websites:

www.cfz.org.uk
www.weirdweekend.org

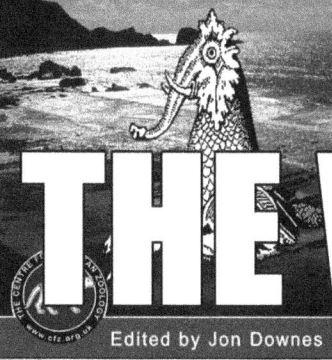

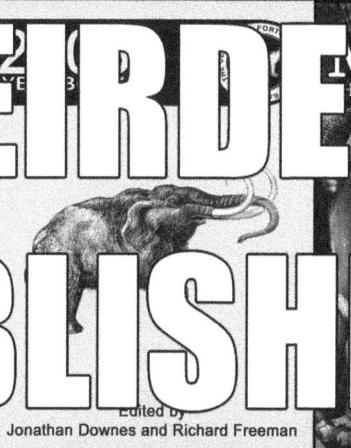

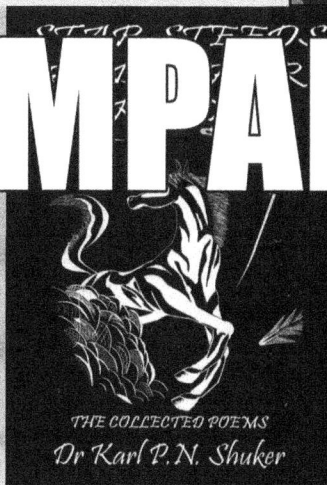

THE WORLD'S WEIRDEST PUBLISHING COMPANY

HOW TO START A PUBLISHING EMPIRE

Unlike most mainstream publishers, we have a non-commercial remit, and our mission statement claims that "we publish books because they deserve to be published, not because we think that we can make money out of them". Our motto is the Latin Tag *Pro bona causa facimus* (we do it for good reason), a slogan taken from a children's book *The Case of the Silver Egg* by the late Desmond Skirrow.

WIKIPEDIA: "The first book published was in 1988. *Take this Brother may it Serve you Well* was a guide to Beatles bootlegs by Jonathan Downes. It sold quite well, but was hampered by very poor production values, being photocopied, and held together by a plastic clip binder.

In 1988 A5 clip binders were hard to get hold of, so the publishers took A4 binders and cut them in half with a hacksaw. It now reaches surprisingly high prices second hand.

The production quality improved slightly over the years, and after 1999 all the books produced were ringbound with laminated colour covers. In 2004, however, they signed an agreement with Lightning Source, and all books are now produced perfect bound, with full colour covers."

Until 2010 all our books, the majority of which are/were on the subject of mystery animals and allied disciplines, were published by `CFZ Press`, the publishing arm of the Centre for Fortean Zoology (CFZ), and we urged our readers and followers to draw a discreet veil over the books that we published that were completely off topic to the CFZ.

However, in 2010 we decided that enough was enough and launched a second imprint, `Fortean Words` which aims to cover a wide range of non animal-related esoteric subjects. Other imprints will be launched as and when we feel like it, however the basic ethos of the company remains the same: Our job is to publish books and magazines that we feel are worth publishing, whether or not they are going to sell. Money is, after all - as my dear old Mama once told me - a rather vulgar subject, and she would be rolling in her grave if she thought that her eldest son was somehow in `trade`.

Luckily, so far our tastes have turned out not to be that rarified after all, and we have sold far more books than anyone ever thought that we would, so there is a moral in there somewhere…

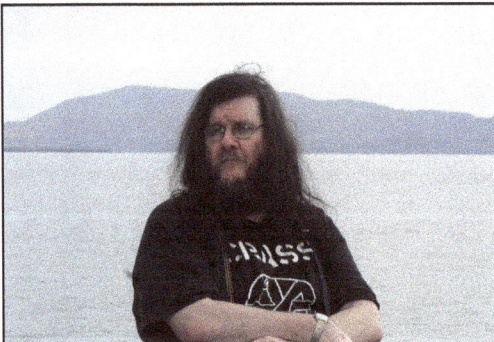

Jon Downes,
Woolsery, North Devon
July 2010

CFZ PRESS

CFZ Press is our flagship imprint, featuring a wide range of intelligently written and lavishly illustrated books on cryptozoology and the quirkier aspects of Natural History.

CFZ Classics is a new venture for us. There are many seminal works that are either unavailable today, or not available with the production values which we would like to see. So, following the old adage that if you want to get something done do it yourself, this is exactly what we have done.

Desiderius Erasmus Roterodamus (b. October 18th 1466, d. July 2nd 1536) said: "When I have a little money, I buy books; and if I have any left, I buy food and clothes," and we are much the same. Only, we are in the lucky position of being able to share our books with the wider world. CFZ Classics is a conduit through which we cannot just re-issue titles which we feel still have much to offer the cryptozoological and Fortean research communities of the 21st Century, but we are adding footnotes, supplementary essays, and other material where we deem it appropriate.

http://www.cfzpublishing.co.uk/

Fortean Words is a new venture for us. The F in CFZ stands for "Fortean", after the pioneering researcher into anomalous phenomena, Charles Fort. Our Fortean Words imprint covers a whole spectrum of arcane subjects from UFOs and the paranormal to folklore and urban legends. Our authors include such Fortean luminaries as Nick Redfern, Andy Roberts, and Paul Screeton. . New authors tackling new subjects will always be encouraged, and we hope that our books will continue to be as groundbreaking and popular as ever.

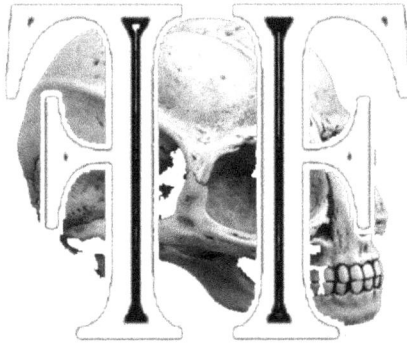

Just before Christmas 2011, we launched our third imprint, this time dedicated to - let's see if you guessed it from the title - fictional books with a Fortean or cryptozoological theme. We have published a few fictional books in the past, but now think that because of our rising reputation as publishers of quality Forteana, that a dedicated fiction imprint was the order of the day.

http://www.cfzpublishing.co.uk/

www.ingramcontent.com/pod-product-compliance
Lightning Source LLC
Chambersburg PA
CBHW051433270326
41935CB00018B/1807